Interviews with Beekeepers

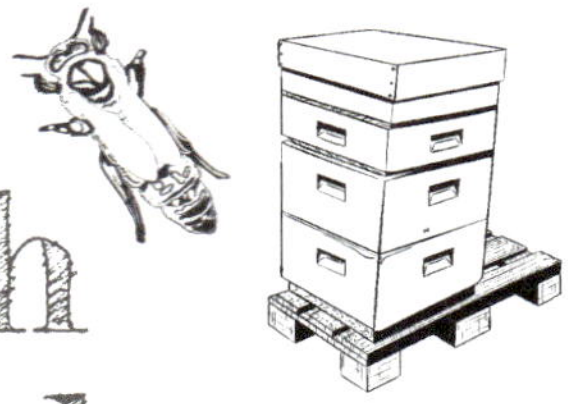

Steve Donohue
30 Mar 2020

Enjoy!

Interviews

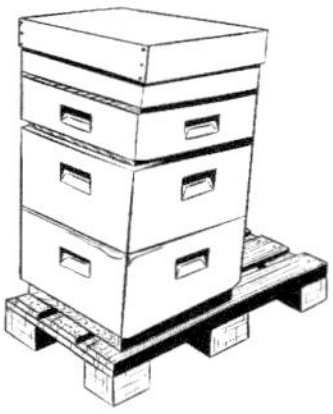

with Beekeepers

Steve Donohoe

First published in 2020 by
ZunTold

www.zuntold.com

Cover designed by Isla Bousfield-Donohoe

British Library Cataloguing-in-Publication data
A catalogue record for this book is available from the British Library

ISBN: 978-1-9162042-5-6

Printed and bound in the UK by Print on Demand Worldwide

Why this book is different

I don't have to imagine what a beekeeper would love to read because I am one.

After seven years as a beekeeper, I was at the stage where I had read most of the books written for beginners and wanted to learn more about these amazing insects and the art and science that makes up good beekeeping. I didn't want to revisit yet again the things churned out for beginners which tell me about the three castes, their life cycles, the various types of hives and equipment; I was ready for more.

My idea was a simple one: interview some of the best beekeepers in the world and let their experience and wisdom be accessible to the rest of us. I hope and trust that the interviewees I have selected, whom I believe are fascinating and fun, have something to offer all beekeepers with an open mind and a desire to learn.

Increasingly, beginners are finding their way to online beekeeping forums for inspiration and guidance. Such a route is strewn with risks because you have no way of knowing for sure that the keyboard warriors who hand down advice know what they are talking about. Any forum has its cliques, its trolls, its zealots for one thing or another, which can be unsettling to the unwary "newb." One example is regional variation. Beekeeping is different from region to region, so what works in one place may be hopeless in another. As beekeepers we should spread our net wide in the quest for understanding, but we must be pragmatic when applying lessons learned to our bees in our part of the world and consider the climate, the plant life and variations of all kinds. Equally, regional differences should not be an excuse for burying heads in the sand and refusing to adapt and learn; the honeybee is an adaptable survivor and beekeepers must demonstrate similar qualities.

This book does not include hobby beekeepers with a few colonies at the bottom of their garden. My focus has been commercially minded folk who make a living from their craft. A bee farmer who makes his or her living from bees, and has consistently done so for decades, is very likely to have developed practices and behaviours that work. I was curious to understand their journey, the mistakes they made and the lessons learned; to me, these things were like gold dust. Despite what many may presume, the commercial beekeepers that so generously gave me their time, are lovers of bees and wildlife. They have no desire to stress their bees because ultimately if you look after the bees, they look after you.

For anyone who is interested, my time with honeybees began by chance. I stood awkwardly at a house party that somehow, I'd been persuaded to attend, wringing my hands and trying to dream up something interesting to say to the other guests. I bumped into a similarly introverted chap called Richard, who turned out to be a farmer, and somehow the subject of honeybees came up. I said I would keep bees if I had somewhere to put them and he said that I could use a corner of one of his fields. That gave me the incentive to study at evening classes over the winter, and by June the following year, my first two hives were in place. I have Fay Hine to thank for me becoming a beekeeper because it was her party, and Richard the farmer was her friend.

I want to say something about the therapeutic effect keeping bees had on my often-racing mind. At the time I began my bee-keeping journey, I was a director and joint-owner of a company that my wife and I were trying to grow. Our long-term plan was to one day sell the business so that we would have financial freedom, but for many years it was an endless battle of problem-solving and dealing with issues. Staff issues, customer issues, money issues, technology issues; these were the things that occupied my head day and night. When you run a small business, and all of your wealth and prospects are tied up in it, the stress and anxiety can sometimes outweigh the sense of purpose and fun that you felt at the beginning. As a younger man, I had drowned such emotions in oceans of alcohol, but I learned many years ago that for me alcohol was no longer an option. That is a road that I hope never to travel again. Then along came honey bees.

There are endless wonders in and around a beehive. As a city dweller, the opportunity to escape to the glorious Cheshire countryside and tend to my bees has always been a source of spiritual nourishment. I grew up in the New Forest and remember that as children we would hurl dried lumps of horse shit at each other, just for fun. There is wildlife in cities; I have frogs in my garden, dragonflies, moths, bumblebees and sometimes foxes, but it's not the same as living closer to nature. I find that when I inspect a colony of bees, my mind clears and I get lost in a world hidden from non-beekeepers; a world of wax and pollen and propolis. My therapy is bees.

Contents

Murray McGregor, Coupar Angus, Scotland

'...when it comes down to a philosophy about bee keeping, it's that the best way to keep bees is the way that the bees would like to be kept.'
(Murray McGregor)

Introduction to Murray

I had heard of Murray McGregor long before I met him. When time allows he posts on Twitter and occasionally on the UK beekeeping forum. Such is his in-depth knowledge, based on experience rather than just theory, that most disagreements on the forum end as soon as Murray posts. It still surprises me that the bee farmer with the most hives in the UK comes from Perthshire, Scotland. Denrosa Apiaries, owned by Murray and by his father before him, had about 3,500 colonies at the time of my visit. Despite many happy holidays to Argyle and Bute and a love of the beautiful Scottish Highlands, I must have a slightly tainted Southern view of our Northern neighbours. How can bees do well up there in the cold and wet? It turns out that the Perthshire climate is perfectly capable of supporting Murray's operation. It is home to some thriving fruit farms as well as arable crops, but the main goal for Denrosa is to collect as much honey as possible from the prolific heather that covers the hills in late summer.

In common with many of my interviewees, it took a long time to persuade Murray to give up his precious time to speak with me. Eventually, we settled on the last weekend of November 2017. As I crossed the border it began to snow heavily, and I had that sinking, "Oh no, this is never going to happen," feeling. After a few miles, it cleared up, and by the time I arrived at my hotel in Alyth, the weather was cold but clear. The Lands of Loyal Hotel was gearing up for Christmas; a huge tree had materialised in the lobby, but the decorating was yet to start.

That evening I dined out with Murray in Blairgowrie and made plans for the following day. He was keen to drive me to several of his apiaries in the area so that I could take photographs and ask questions along the way. We started our meal with a generous bowl of something called Cullen skink, which is a delicious creamy fish soup. I had never heard of it before but will certainly look for it on any menu from now on. I don't know if the term, "gentle giant," is quite the correct description of Murray but that was my initial impression. He is an imposing figure with an unexpectedly soft, soothing Perthshire voice. I suspect, however, that if I was an errant employee, I might see a less gentle side to the man; he does possess a piercing gaze.

The next day I was given a tour of some of Murray's apiaries in his little red Fiat van. These were his over-wintering sites, carefully placed away from view along the valleys of the Rivers Isla and Tay. One memorable spot was on a fruit farm. Murray had his hives directly beneath an electricity pylon, to show that the old myths about bees hating electric cables are just that. In my interview with him, later on, I asked about choosing apiary sites because I know that it is something of a passion of his.

After a break for a bacon sandwich, we returned to base, and I met Jolanta Modliszewska. She runs the queen rearing side of the business, which is vital in bee farming; good queens are the lifeblood of the whole operation. I interviewed Jolanta in the queen rearing shed in the land outside Murray's house, and then, later on, I interviewed Murray himself.

Talking with Jolanta and Murray

Queen rearing

Steve: Do you want to sit down?

Jolanta: No, it's fine.

Steve: Murray said that you are an important part of his business.

Jolanta: Yes

Steve: I've been interviewing other bee farmers and its always the queen selection and queen breeding that is the critical part of their business, because that determines their success, doesn't it? So how did it start for you?

Jolanta: Actually, I worked with Murray for a few seasons as a beekeeper. It was getting a bit too hard for me, because I love bees so much and I wanted to expand my knowledge. I thought that queen rearing would be good for me so I pushed Murray, and he said that it had been his goal for many years to do that.

Steve: Right

Jolanta: Four years ago he built this shed and I started off with 150 mating boxes, and was quite successful in the first year. The first year was a trial, then we slowly started to expand each year. We are now up to 900 mating boxes.

Steve: Do you mind explaining how you go about it?

Jolanta: How the whole system works?

Steve: Yes

Jolanta: OK. I have forty grafting frames and each frame holds 28 queen cups. My system is that at the beginning of May I ask Murray about bringing here some great colonies from out in the fields. I usually do the first grafts around 10th May. Last year I had four starters, and for each starter I have two finishers, because I can only put a maximum of 15 cells into each, to get nice big queen cells.

Steve: OK

Jolanta: I have my best queens in smaller boxes so they don't lay many eggs and can live quite a long time.

Steve: How often does a new breeder queen come in?

Jolanta: It's hard to say actually.

Steve: Is it just when you get a good one?

Jolanta: Yeah. I'm quite choosy, because we have 3,000 colonies, so it has to really be great. There is a huge European Foul Brood (EFB) problem in Scotland so I always check for that as well. Mostly, and Murray will tell you this, the black bees (Amm) are the ones who have this problem. We don't really breed from the black bees but if it's really great it can come here. Of course, some queens come from abroad as well. Murray buys them; I think I have some from Cyprus and Italy. I learned how to do queen rearing in Cyprus.

Steve: From Roger White?

Jolanta: Yes

Steve: How was that?

Jolanta: Yes, I had a nice time.

Steve: Did he teach you about grafting?

Jolanta: Yes.

Steve: I've only done it once and it wasn't very successful, but I have something wrong with my eyes.

Jolanta: OK

Steve: I've been told that I need to get a jeweller's loupe, or something?

Jolanta: I have something like that – see this one here

Steve: Oh, that's great. Where did you get it?

Jolanta: From Amazon I think. It was a gift from somebody.

Steve: It works well?

Jolanta: I don't use it. I still have good eyes so I don't need it.

What I use for grafting is the Chinese tool, but the wooden one. I don't like the plastic ones; the wooden (bamboo) ones are better, and I use a headlamp. I close the shutter so it's dark and keep a moist towel on the frame so larvae don't dry out. I keep the frame in the starter for two days. Each day I do four frames, so every day I'm doing grafting.

Steve: Is the starter colony just a big colony?

Jolanta: Yes. Not all colonies are good starters, so sometimes I change them. If it's wrong I will send it away.

Steve: How does it work, are they queenless?

Jolanta: Oh no, all starters have a queen.

Steve: With a queen excluder?

Jolanta: Yes. I keep my good queens on five frames in a nuc, then a queen

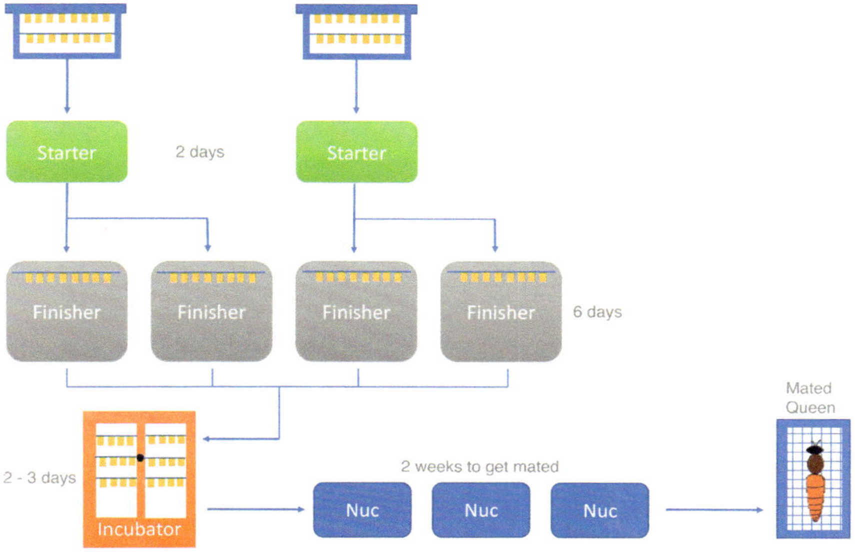

excluder on top, then another nuc on top of that with five frames. The frame of grafts goes in the top to get started.

Steve: When you do the graft is it into dry plastic cells or do you put anything in.

Jolanta: Sometimes when it is very hot it's good to put royal jelly in the cups first, mixed with water fifty/fifty. My teacher does that all the time in Cyprus because it's much hotter there. Here it is not usually necessary; it is quite mild but not hot summers.

Steve: To make sure that you have larvae of the right age is there some method for that or do you just find the right frame?

Jolanta: No, I just find a frame where the eggs have just become larvae.

Murray: She takes a frame that has been laid up, and she takes the larvae from the transit zone where they are just changing from eggs to larvae.

Steve: So, you take the really young larvae?

Jolanta: Yeah

Steve: Did you find that difficult at first?

Jolanta: No, not really.

Steve: You just go under it with the reed?

Jolanta: Yes. You go underneath the outside of the "c" not the inside. Never the end or the inside of the "c". [referring to the c-shape of a young larva]

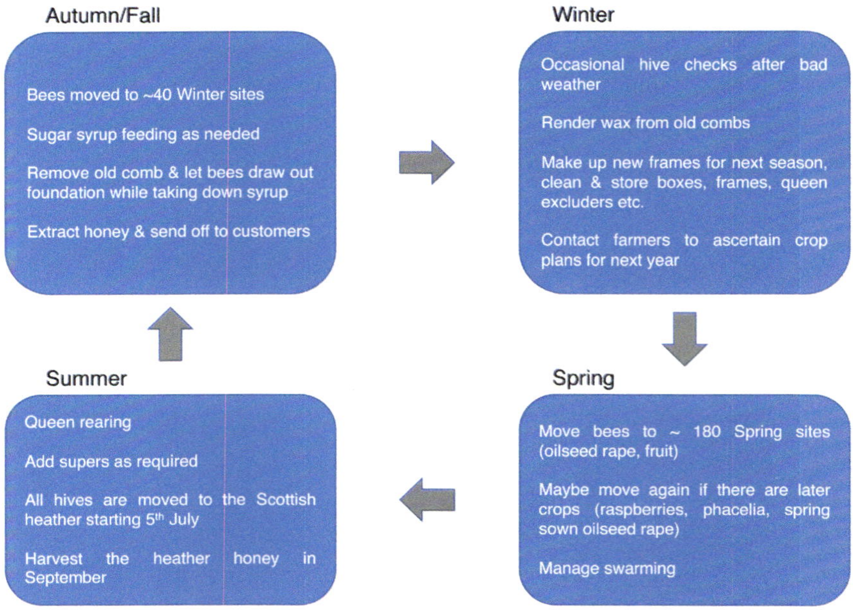
Autumn/Fall
Bees moved to ~40 Winter sites
Sugar syrup feeding as needed
Remove old comb & let bees draw out foundation while taking down syrup
Extract honey & send off to customers
Winter
Occasional hive checks after bad weather
Render wax from old combs
Make up new frames for next season, clean & store boxes, frames, queen excluders etc.
Contact farmers to ascertain crop plans for next year
Spring
Move bees to ~ 180 Spring sites (oilseed rape, fruit)
Maybe move again if there are later crops (raspberries, phacelia, spring sown oilseed rape)
Manage swarming
Summer
Queen rearing
Add supers as required
All hives are moved to the Scottish heather starting 5th July
Harvest the heather honey in September

Steve: OK. Then you pop them in the cups and put the frame in the starter.

Jolanta: Yes, just for two days, then afterwards into the finisher, which also has a queen.

Murray: Everything is queen right. You don't need to replenish anything when they are kept queen right.

Steve: I was going to ask about that.

Murray: You don't need to bring in any fresh bees or fresh brood; they just keep going all season when they are queen right.

Steve: They'll happily do that many, will they?

Murray: In the right conditions they'll do practically the whole lot. It's very rarely that you get 100% (of queen cells from grafts)

Jolanta: Actually, it happened once last year that I lost just one out of the whole frame, but I've never had 100%

Steve: I guess anything over…

Murray: More than half is perfectly good

Jolanta: So when I get a lot I have to split them into two finishers

Murray: That is so they all get well fed

Steve: Do you feed the finishers with syrup or pollen or anything?

Jolanta: If the weather is fine and there is lots of pollen and nectar coming in, that is the best, I think. Natural sources of food are better. When it's not like that I do feed them, yes.

Murray: But one thing you never have to do here is feed pollen; there are no pollen dearths here. They can always find pollen.

Steve: So, as far as timing goes…you graft every day, you said?

Jolanta: Yes, every day I do four frames

Steve: Then once they are sealed?

Jolanta: They go to the incubator.

Steve: Always?

Jolanta: Yes. Then for the first filling of mating boxes I wait for virgins to hatch. I don't put a queen cell in for the first time.

Steve: Why is that?

Murray: It's a far better acceptance rate. If you are starting from scratch there is a far better acceptance rate with virgin queens than queen cells.

Jolanta: Yes. Later, when I have taken away the mated queen, from then on I put in a queen cell, not a virgin.

Steve: How many days do the cells stay in the incubator?

Jolanta: I think they go in at eleven days, so they are in there for three or four days.

Steve: Then you've got to make sure that you have somewhere to put them!

Jolanta: Yes. I say to Murray, "I have fifty virgins – I need packages!" [laughs]. I do not want to keep virgins for longer than two days maximum. I give them a little syrup in their roller cages.

Steve: Then Murray turns up with this thing – is that right? [pointing to ventilated plastic box for packages of bees]

Murray: Yeah, or two or four or six of them, depending on how many she's asked for.

Jolanta: One day I had 80 virgins that needed bees. We can get about 12 mating boxes populated from one package of bees.

Steve: Is it those little tiny mating nucs? What are they called?

Murray: You get micro nucs, apideas…we use the slightly bigger ones called the Kieler, although we do have some apideas. That's a kieler over there – it's just a top bar hive; you use top bars, not frames.

Steve: No drawn comb in there?

Murray: No, foundation is best.

Jolanta: Then I put them somewhere cool and dark for two days

Steve: OK. Is the mating area around here then?

Jolanta: This is one area. The second one is at Coupar Angus. There are 500 mating nucs here and 400 in Coupar Angus. I have a better mating rate at the Coupar Angus site; it's a more sheltered place.

Murray: It's an old neglected orchard alongside a walled garden. You've got a high wall on one side and then high trees around three sides, so it's very calm and sheltered. The bees fly there in conditions that they don't fly here.

Steve: Is the first time they go out into mating boxes late May/early June?

Jolanta: Middle of May, when I start grafting at the beginning of May.

Steve: Does it have to be a certain temperature for them to go on a mating flight?

Murray: Theory says 20 Deg C but many queens in Scotland mate below that temperature.

Steve: Otherwise they would never mate at all! [laughs]

Jolanta: Yes. I think it's less than 20 Deg C for sure.

Steve: Then you've got a mated queen in a little box.

Jolanta: Yeah.

Steve: Is your part done then? Is that still you or is it passed on to somebody else once she's mated?

Jolanta: Once the queens are mated, from around June, it can get busy so I get help then.

Murray: I think what you were getting at is when does the responsibility transfer from her to the bee team. She has to catch them, mark them, cage them and then a box of cages gets handed over to the next step in the chain. Also, if she's got time, she'll take the cages out and put them in the hives as well.

Steve: Do they always go to nuclei first?

Murray: No. Just as required. They go into nuclei or full hives or they go away in the post to a customer.

Steve: And you are basically at it for the whole summer?

Jolanta: Yes.

Steve: What is your favourite bit?

Jolanta: Catching queens! Definitely [laughs] I like to see my babies.

Steve: To see how they have done?

Jolanta: Yeah, that's the best.

Steve: When you have caught the queen, what goes in next, another virgin?

Murray: A queen cell wrapped in foil, to stop them attacking it.

Jolanta: That's my system; the second time I always go with a queen cell. If I put a virgin in then it would fail. One year I tried with virgins and had zero success.

Steve: Do you wrap the whole queen cell in foil?

Murray: You just leave the tip

Steve: Then they just leave it alone?

Jolanta: Yep. Sometimes it doesn't work but I think it's 80% acceptance with this method.

Steve: When you catch the queen do you clip one wing?

Jolanta: I don't clip them. I just mark them.

Murray: We don't clip queens in the year that they are born. If a customer who lacks confidence asks us to clip them we will do it, but normally we won't do that until the next spring.

Steve: And you were saying that with marking you've got two dots?

Murray: No, we've got parallel colour codes. The 2017 colour is yellow but we use orange. Next year it's red but we use pink. That's just for the queens born in this unit. Any queens raised in the field or bought in from elsewhere all have the international colours.

Steve: OK. Is that how you know they are yours?

Murray: Yes. We know one of our own queens.

Steve: How many breeder queens are there in nucs around here?

Murray: Some are in nucs and some are in big boxes but she'll generally have 25 lines or something, between drone mothers and breeders.

Jolanta: I use some just for starters. It's hard to say how many because it changes from year to year.

Steve: What about drones? You bring in good colonies but do you do anything special?

Murray: In both mating sites we have apiaries close by, then we put colonies there selected because they are so good. The colonies here, and we've got twenty odd lines, also throw off plenty of drones. The chances of inbreeding are low.

Steve: I'm reading a book about mating where they were saying that the queens fly much further away to mate than her drones do to prevent inbreeding.

Murray: They also fly at different times of the day in some cases. Also, when you've got a lot of virgin queens you get a lot of incoming drones; they are attracted to the pheromones.

Steve: Do you happen to know where the drone congregation areas are? No. It doesn't really matter does it?

Murray: It doesn't matter.

Jolanta: Sometimes you can hear them – a big cloud of drones.

Murray: They go out, they mate, they come back and it doesn't greatly matter to us where the DCA is.

Steve: Do you ever see them go?

Jolanta: The virgins? Yes, I sometimes see that.

Steve: They fly off with a cloud of bees don't they?

Murray: Not really with the mini nucs. They are almost on their own. There will be a few others going out but whether or not they travel as a group, you can't tell. There are hundreds of boxes around and the air is a maelstrom of bees sometimes.

Steve: What percentage get mated?

Jolanta: I think 60–70% maybe but it depends on the weather.

Murray: Yeah.

Steve: When it fails is the queen just gone?

Jolanta: Sometimes, if it's hot, I have problems with them absconding in apideas. It's just empty, and there are small swarms hanging around.

Steve: Otherwise it's just a drone layer?

Jolanta: Yes, a drone layer, or something wrong with the queen. Maybe she has a bad wing and didn't get mated because of that. This year I passed over 1,000 queens mated so we get more and more. I'm happy with this. We have such a short season.

Steve: How do you keep track of the different lines?

Jolanta: I have a numbering system.

Murray: She uses this notepad to record everything.

Jolanta: Yes. Lots of dates and numbers...date of grafting and number of mating box and queen lines...

Steve: Do you review all of these notes to see what went well?

Murray: Not particularly. You just know, without doing anything too scientifically. There's a whole load of successful matings there, just one "F". That line, queen J21; they are really nice queens actually. [Murray pointing to Jolanta's notes]

Steve: So you know them by their numbers?

Jolanta: Yes, and I have my favourites. [laughs] Of course, Murray sees the results of my babies, so he can tell me if a line is not good or if he sees problems.

Murray: If that happens they get put back in a normal hive and sent away to work.

Jolanta: Yes.

Steve: How many get sold, about half?

Murray: No, not even that. The main function of the unit is to make queens for ourselves but we are getting more and more orders, and more and more big orders.

Steve: Would you like to have more sales of queens?

Murray: This is not a unit of finite size. It can grow to fit demand, and Jolanta can have a full time assistant if she needs it.

Steve: Jolanta, would you like Murray to grow the queen rearing side?

Jolanta: Yeah, probably

Murray: It's her domain, and she's becoming well known for the quality of her work

Steve: That's great isn't it?

Jolanta: Yeah [laughs]. Thank you.

Murray: The Scottish bee inspector and the woman from the science institute were saying only last week that they think this is a great project.

Steve: Do you keep in touch with Roger?

Jolanta: Yeah, I have his number

Murray: She's new on the Buckfast group. I'm not.

Jolanta: Yes, I'm on the Buckfast group. I will not say that Buckfast are my favourite but...

Murray: We tend to have something of a favourite here, the Carnies, because they seem to be a bit more suited to our climate than the Buckfasts. The thing with Buckfast is, there's no such thing as "a Buckfast". Every breeder's Buckfasts are different. Roger's bees are very nice; they build up and get big strong colonies, but they under-produce in our climate.

Steve: Well, they're from Cyprus

Murray: Not really. He breeds them in Cyprus but they are actually northern origin stock. A lot of the queens from Southern Europe and from warmer climates are actually from stock selected in the North, taken to the South to get the early season, but the genetics are northern. One thing you don't want is the "Cyprus Bee" [Apis mellifera cypria], because that is an aggressive bee. Roger has to work hard to keep his stock pure by using instrumental insemination and drone flooding for his open mated queens.

Steve: For someone like me, an amateur with a few hives, is it still worth trying to raise queens? It seems to be worth trying.

Murray: There's one danger; too many from one line and you can lose your vigour from inbreeding. We don't like any more than about fifty daughters from any breeder going out into our main unit for mating, so that we don't get the gene base narrowing down.

Jolanta: Yes, agreed

Murray: Amateurs who do insemination and selection from their best stock, even if they have 10–12 hives and they do three from one, they are actually going more inbred than we would ever like to go.

Steve: Right. So, you would say for someone like me with 10 hives that it would be better to find a local queen breeder who is doing what you are doing?

Murray: The thing is, you don't know what mother the breeder is choosing from; they may all be from the one mother if you bought queens off them. You don't know and you've got to be careful with that. You are probably better with a range of sources to try to keep the diversity. But then breeders sell their breeder queens to each other as well. It's worth asking what the mother lines are if they'll tell you, because they may all come from the same place. Keld Branstrup in Denmark supplies breeder queens to quite a number of people even in the UK. You may think you are buying two lines from two different people, but it may all lead back to one mother queen.

Steve: Are there any things that you found really difficult at first, or any tips or tricks that you have learned, Jolanta?

Jolanta: Actually I have to say that I have not really found anything difficult [laughs]

Steve: You just get on with it and take it all in your stride.

Jolanta: Maybe Murray can say something? I think I maybe care a bit too much

Murray: You care a lot, and you are too self-critical, and worry too much about things that you don't need to worry about

Steve: Have you ever had any disasters?

Jolanta: I was quite upset last year when I had 100 nice virgins and all failed to mate. I was really upset about that.

Steve: All were what, sorry?

Murray: All failed

Steve: Oh. Just bad weather or something?

Jolanta: No no. I just put them into established mating boxes, and that is why I don't do this now; this was my lesson

Steve: Oh, I see.

Jolanta: I think I will cry when number seven dies, because she has been with me from the beginning, from my first year of queen rearing. I am getting like an old lady! [laughs]

Steve: Can you tell that "it's going to be a good day today"?

Murray: Sometimes in the morning you can say, "a lot of drones will die today!" [laughs]

Jolanta: I am quite happy when after two days of bad weather the sun comes out so that they can mate. In bad weather I am always worried that they will not be able to mate. It is a relief when we get the sun.

Steve: How long before it is too late for mating? Is it three weeks?

Jolanta: I think three weeks is a bit too long. I wait for two weeks.

Steve: If they are not mated then you get rid of them?

Jolanta: Yes

Murray: Turf them out and start again. Once the bees get too old the only way is to turf them out and make up new mating nucs.

Steve: It's farming isn't it? Some things you control, some you don't.

Murray: Yes

Steve: OK. Well thank you. Very interesting.

Jolanta: Thank you

Steve: You are recommending the wooden Chinese grafting tool?

Murray: The one made from bamboo, yes. They are very hard to find.

Jolanta: Those ones are the best for me.

Steve: OK, thank you very much.

The following year Jolanta's beloved breeder queen J7 expired. Nothing lasts forever. After fortifying myself with a cup of tea it was time to formally interview the man himself.

Murray's story

Steve: How did you start?

Murray: My father started keeping bees in 1950, so as a young child I was brought up with bees in the family. They were with us all of the time. My mother used to extract the honey while my father did other things. I would be with her in the pram even, when I was tiny, but I started actually going to the bees with my father when I was eight. There was a little barrow that I used to carry supers in. I'd take it over a little bridge to the apiary in the evening when Dad came home from work.

When I got home from school, if I knew it was going to be an evening with the bees my mother had terrible trouble trying to get me to do my homework, because all I wanted to do was go out to the bees with Dad when he came in at five o'clock.

Steve: You weren't scared of them or anything?

Murray: No, not especially. I mean I didn't like being stung, I never did, and I still don't, but it's just a fact of life, it happens.

Steve: That was from eight years old...

Murray: Yes, then I got my first bees when I was fourteen years old. I used to pick raspberries in the summer for money, along with many young people from Coupar-Angus. It was strawberries and raspberries in the summer, potatoes in the autumn, and once I had the money my father took me to a town near Arbroath, to somebody who was selling bees, and the money was duly invested in my first hives.

Steve: You obviously had a good teacher in your Dad.

Murray: Yes, he was a good beekeeper and he had guidance from good beekeepers, so we were surrounded by the bee culture from an early age. He didn't just do bees, he was an electrician and did some market gardening just to try to make ends meet. He grew dahlias and chrysanthemums and in the mornings before school we'd be cutting them to be taken to the florist. My mother had an acre of blackcurrants which we used to pick and sell too. Then there were the bees.

Steve: As you got going with bees and were learning from your Dad...

Murray: Well, only for about three years, because at seventeen years of age I made the decision of infinite wisdom to leave school and join the merchant navy. I joined Shell as an officer cadet.

Steve: Oh right, OK.

Murray: My bees then were just incorporated into my fathers, which was a far larger operation. He had 300 hives and I had 12, so it just got absorbed.

Steve: So that was the end of beekeeping for a while?

Murray: Yes, but I realised from quite early on when I was at sea that I'd like to come back and take over the bees, and run it after my father. I liked bees, I liked the countryside and being in tune with nature. Working with the rhythm of the seasons gave me a good bit of pleasure. I was always a bit of a dreamer.

Steve: Other than your father did you have any beekeeping heroes as it were, or people that you read up on, or were inspired by?

Murray: I would have to admit, possibly to my shame, that I had never actually read a beekeeping book from cover to cover. Even today the number I have read I could number on the fingers of one hand. I didn't have any other beekeeping heroes apart from my father. There was no particular beekeeper who I thought was wonderful, but there were a couple I thought were disastrous. I learned mostly from my father and once I was back working as a professional beekeeper he remained my best advisor, friend, and confidante until he died a couple of years ago.

Steve: That's nice isn't it, to have that bond?

Murray: Yes. He was so interested in what was going on with the bees that if I came in to see him and anyone else was there with him they knew right away that his time for them was over, because even at 95 years old he would say, "I want to hear about the bees."

Advice on becoming commercial

Steve: One of the things that I always ask is about people (and you have referred to dreamers) who think that having successfully kept 10 hives they can scale up to hundreds and go into business. I just wondered if you have any advice for those people? To go into business as a honey farmer what do you need to keep in mind before throwing your money at it?

Murray: You have to bear in mind that it is a very inconsistent profession. You'll get years when you think it's easy and you'll make money and do very well, but you'll get at least as many years when you make a loss. It is not an easy thing to do long term. It is best for people who have already got significant money behind them and they can afford to take a few seasons of losses. The best plan is to go from being an amateur beekeeper to part-time professional; keep your job and do your bees at evenings and weekends. Once you've ironed out the errors and you are starting to do consistently better than the average Joe around you, then that's the time to jump to full time professional. You must always be aware of the risks.

Steve: If somebody near you had such aspirations, I guess you could argue that they are future competition, could they come and work with you for a few years to learn the ropes? Would that be a good path?

Murray: We generally don't take people on in that kind of way. You have to accept in life that if somebody works for you and they decide to go off on their own and run their own business that's just a fact of life, and you just have to accept it, and feel good that you played a part in getting them there. You can't resent it because you can become obsessed with inconsequential matters if you do. We would not particularly take anybody on who at the outset declared that their aim was to go into full time business on our patch.

Steve: No, that wouldn't make sense would it.

Murray: If they stated initially that their intention was to go into business we wouldn't hire them. We train people on the job with us in the hope that they're going to work with us for several years.

Steve: Can you say roughly how many hours you spend per week working?

Murray: It's probably somewhere between 28–29 hours per day! Seriously though, in a field like this, the distinction between work and leisure is very blurred. This is a life rather than a job. You live the bees. The bees are your life, and you can quite often find yourself working with bees all day and then, what to do in the evening? Look at more bees! Because it's a pleasure as well as a job, you can spend a lot of your time doing things that are not particularly constructive in terms of achieving some work target, but they are actually good for your soul. You go around and just look at things and smell things and see what's happening – it's not work per se, but you're still at the bees. So, to distinguish between the number of hours that are actual work from those that are just pottering around is quite hard. There will be weeks where, if you took a strict line on how much was work, then some weeks I'll do nothing, but on other weeks I'll easily do 100 hours.

Revenue splits

Steve: Just thinking about your revenue, you've got honey, split into normal honey and comb honey, then you have nucs and queens…how does that break down?

Murray: Well, 80% of our income is from heather honey, as a ball park. The rest is made up of blossom honey, nuclei and queens, and trading in package bees. Roughly speaking, as a general average, about 10% of the heather honey revenue will come from comb honey. The rest is sold in bulk in barrels.

Steve: Does it ever happen that the heather doesn't flower?

Murray: You occasionally get seasons where the heather doesn't work. After the foul brood outbreak in 2009, up to 2016, we had five poor years (meaning below average years) in that period. We've now just had two above average years, one after the other, but that doesn't quite make up for the shortfall. There are years where the heather doesn't flower well, or the flowers have very little nectar, because it hasn't had enough sun on its back in the lead up to the flowering period. If it doesn't get the sun at that time the plant doesn't make the sugars that become its nectar.

Steve: Can you tell when that's coming?

Murray: Not especially, no. There are years where you think it fits the bill for a failure and they're not, and there are years you think are going to be a Klondike, and they're not. You just have to move them all to the heather anyway and see what happens.

Worst years and best years

Steve: Can you think of a particularly horrific bad year, like the worst year ever?

Murray: My worst year was 1985, which was followed up by 1986 which was almost as bad.

Steve: Not really what you want, is it?

Murray: No. We had a disastrous year; 1985 was a terrible summer. A terrible summer actually hits you with a double whammy, because there was very little blossom honey, then it rained all through the heather. It rained almost every day from the end of June to the end of October. There was flooding, no heather honey and the bees stopped raising young bees, which was what made it so very serious. We had very little honey, and what happened with so few young bees is that we had a huge spate of winter losses. The hives that survived winter and were alive in early 1986 were under sized and took much of the year to recover. So, 1986 was just as bad a honey year as 1985, but was caused by a combination of it being another poor summer and the lack of bees early on. It took two to three years to recover from 1985.

Steve: When did you start?

Murray: My first full season was 1982

Steve: Right, so this disaster was early on in your career.

Murray: Yes. 1982 was OK, 1983 was a poor one and the heather flowered very late, then we got a bumper year in 1984 followed by a catastrophe in 1985.

Steve: What about, conversely, any amazingly good years?

Murray: We have had a couple of real crackers. My father always was on about 1955. It was his best year and the year I was born. He had a phenomenal amount of honey per hive. It was all early season honey before the heather. We got a season in the 1990s, I think it was '92 or '93, which was very, very good, and the only time ever that my average per hive beat my father's best year in 1955. It was quite a golden moment. We have had some good crops in the medium past, and these last two years have not been bad, but the year that sticks most in my mind, despite it not being our best one, was 1984 – the one before the catastrophe, because it was a tremendous year for bell heather. Some years we get it, some we don't. My father and I were extracting bell heather honey up to 10 o'clock at night, just to get the boxes free, to get them back to the bees the next day. In some places we were taking off a ton of honey in a day, and taking the boxes back, and 10 days later they were full again.

Steve: It's a good job that you had that year given what was to follow.

Murray: Yes

Steve: Well, now you mention averages and so forth, what would you say is an average honey crop for a production colony?

Murray: Our long-term average? We don't really count the blossom much, but we should be looking at about 10kg for blossom honey. It used to be more like 15kg but clover is more or less an extinct plant round here now. We used to get 15kg to 20kg with the clover and when the raspberries were on, but neither of them are here now, so blossom honey averages have declined to about 10kg per hive.

Heather is our main crop, and our long-term average is about 20kg, 42lbs – 44lbs per hive per year, going back 30 to 40 years. The changes in methodology in recent times means that our average is rising slightly now.

Steve: Is there any particular thing that you think is responsible for that?

Breeding programme

Murray: We've done a fair bit of changing of bee stock. We have a breeding program in place using selected stock that's done well, combined with the fact

that we now try to split everything early and re-unite the splits to the parent hive at the heather, which gives you an extra chance of a crop.

Steve: We were talking earlier about how many hives and how many apiaries you have, and where they are, and you showed me your spreadsheet for that. Can you give me a quick summary of how it all works across the season?

Murray: We regard the start of the season to be in the autumn. It sounds a bit bizarre, but that is the time that you put in place the building blocks which determine what you've got to work with in the spring. We collect all the bees back from the heather to a relatively small number of locations, if you can call 40 sites relatively small. We feed the bees and do our comb renewal so that the bees are in good condition going into winter.

In the springtime we have already collated all the information from the farmers about what crops are growing where, and where the bees are going to go, so we set about dispersing the bees from their wintering sites to a relatively large number of spring sites, which can be 170 to 180 places. They are set up in far smaller groups to avoid overloading the forage in each place. While the oilseed rape is in flower there is abundant forage but as soon as it tails off, if you have too many hives in one place the bees go into decline.

Steve: Right

Murray: So, they are moved out from their winter sites to the oilseed rape. If we have an extra crop to go to like spring sown oilseed rape, or raspberries or perhaps some phacelia or porridge, we will move them onto a second crop because happy bees are working bees. As long as they have something coming in the colonies do well, they stay healthier and have better morale and develop better for the heather. If there is no second crop to move to they stay where they are and we hope they get a bit of clover, something from the trees, maybe Himalayan balsam if it comes along in an early flowering area.

Then from sometime around 5th July we start the heather moving. All of the bees are collected from all of the summer sites.

Steve: Some in England and some in Scotland?

Murray: Yes, we have a unit of bees in England centred on the old Cooperative Farms at Hereford and Cirencester, and they tend to overwinter down there and go to spring crops there, but all hives are moved to the heather. We have four distinct heather areas: Aberdeenshire, which is basically around the valley of the River Dee, West Perthshire, the A9 between Dalwhinnie and Aviemore, and then we also have recently acquired some territory in the Angus Glens, in particular Glen Esk.

They go to the heather to get the heather honey crop, which is the economically crucial one. They are moved there throughout the month of July. At that point we no longer examine the bees, it's just a boxes-on operation – add boxes as required, re-unite splits, and then wait until the end of the first week of September, when we commence the harvest.

Then the whole cycle repeats again. You harvest the heather honey from the bees on the mountains and then they go on the lorry and you bring them home to their winter place, feed them and that leads into the next spring.

Steve: It's a lot of moving of boxes.

Murray: Lots of moving of boxes.

Current issues and controversy in bee keeping

Steve: That's really interesting actually, thanks. Now some out of the blue questions which happen to be next on my list. There are some controversial issues which I sometimes read about, the first being so called "natural beekeeping" – what are your thoughts on that?

Murray: Well, bees have been kept by humans for thousands of years, in a variety of ways. What's fashionable or not fashionable comes and goes in waves. The basic fact is that beekeeping isn't natural for the bees. They don't realise this but it isn't. They would naturally be living in a hole in a tree, or a cave or something like that, so "natural beekeeping" is something of an oxymoron.

I'm not in favour of out and out natural beekeeping because I don't believe it optimises the bees. I don't believe it optimises their health or their productivity. Bees were not kept for any time, apart from the last five years, as some kind of conservation exercise; they were kept because they were useful. Beeswax was used by monasteries for candles and honey was used as a sweetener and a food. This idea of keeping bees just for the sake of it, and that taking honey is bad, which some natural beekeepers believe, is to me just a fashion. It doesn't do the bees any great favours and it doesn't give an incentive for the beekeeper to keep doing so. I worry about the lack of comb replacement, disease control, and I worry about swarming. Some of them keep the bees in remarkably small cramped spaces, which makes swarming inevitable. With swarming, your nice friendly neighbours might not stay friendly for long if they are regularly being besieged by swarms of bees landing on their pergola or whatever.

Steve: Yep

Murray: You have a responsibility not only to the bees, to keep them healthy, but you have a responsibility to your neighbours – that your activities with bees don't impinge on the lives of those around you.

Steve: Presumably treating for varroa would not be natural?

Murray: There are ways to treat for varroa that are considered natural, but the interpretation of what's natural basically depends on the belief structure of the person involved. For example, what a vegan would consider to be natural beekeeping is entirely different to what an organic gardener would consider to be natural. It really depends.

Insulation and ventilation

Steve: Another thing that crops up a lot, mainly between the UK and the USA, is the insulation versus ventilation debate, if there is one.

Murray: I think it's a debate largely between those who want to be fixed in that way. There is no real doubt that in the winter insulation is important and ventilation, especially top ventilation, is anathema to the bees. The bees tell you what they want. If you put a board on top of your hive with a vent to allow for top ventilation the bees gum it up. They don't want it. We've never seen any advantage whatsoever to top ventilation, but a lot of advantage to having insulation.

Steve: I think from what I've heard from discussions over the pond, the idea is that over winter they will get damp without the top ventilation. Do you use solid or open mesh floors by the way?

Murray: We've got both.

Steve: I wonder if that makes a difference?

Murray: I don't think it makes a huge difference but I would suggest that mesh floors are slightly superior to solid floors from what I can see. But, to compare what happens in the UK, with a relatively gentle maritime climate, to what happens say in the Midwest USA or the Canadian prairie provinces, with severe cold, icing up of the front of the hives at the bottom and many problems that we don't encounter – to say in the UK that we know what's best for somebody in North America would be nonsense. They are working in their environment and they've got the best idea of what works for them.

However, like everywhere else, there is a tendency in beekeeping to get bogged down in tradition. So, if traditionally that's the way it's always been done then that's the way it must always be done. In fact, beekeeping is, or should be, like any other progressive craft; open to modern methods so that changes in technology and materials should be exploited to our advantage.

Buy them in or grow your own queens?

Steve: OK. The other one was the importing of queens. You discussed with me importing Italian queens. I guess it is the "buy it in" versus the "grow your own" discussion.

Murray: There's a lot of aspects to it, and a lot of people will say that you are importing all these diseases, exotic pests, and that sort of thing. It's not happening. It never has done, not with legal shipments of queens done the correct way; there are no serious issues with it. To me, it's important that we have access to the stock we need to do the job we need to do. There are other people looking at what we want to do in terms of honey production, and insist that it's wrong, and queen imports should be banned. There are many professional bee farmers in the UK who would be given acute difficulties by such a policy, because they couldn't get their queens early enough in the year.

Our policy is that we're doing a twin track, and you've seen whilst you've been here that we've got our own queen rearing unit rearing a large number of queens each year, and a large number of nuclei. This doesn't get rid of the need for us to have access to imported stock; far from it. To keep the genetic diversity and to keep the right bees we need to buy breeder queens from the best sources, and also to be able to get access to large numbers of queens when we need them. This is not a thing you can plan for in the UK, because the queens sometimes go for many weeks without mating. The flowers are not going to wait until your queens are mated; you do require some kind of reliability so you can plan things.

Having our own bees bred in the North of Italy gives us the option of having mated laying queens from our own stock back in our own hands at least six weeks early. This gives us more flexibility and more chance to get a crop to pay the mortgage.

Neonicotinoids

Steve: OK. Lastly on the controversial things is the question of neonicotinoids – some people are sure they are killing our bees, whereas whenever I talk to a beekeeper they actually quite like taking their bees to the oilseed rape, which is apparently grown from neonic coated seeds.

Murray: Neonic seed coatings have now not been used for at least 2 years, but we never noticed any decline in our bees when they became more common, nor have we noticed a rebound since. It was a non-issue to me. However, it's fair to say that the balance between collateral damage to bees and pesticide use has been pretty healthy for at least the last 30 years. There's not the damage that there used to be in the past. The broad-spectrum insecticides have gone and the bees are generally much the better for it.

My feeling on the neonicotinoid debate is that there may well be something in it in certain areas with certain climates and certain soil types. It's not applicable in our area because we've not seen any negatives at all with it. I did feel rather that we were somewhat hijacked by the movement against the big agriculture companies, and activists were seizing on us as a cause to champion, without there needing to be much evidence that it was a problem. Will I miss neonics? Of course not. The idea that you would like there to be a pesticide floating about in the environment that may or may not be harming your stock is not rational. Yes, we are glad to see them go and in a way it's about what comes next – it could be worse.

Steve: Also, I suppose if it causes farmers to grow alternative crops, could that be a problem?

Murray: It's a problem in the South because there are pests down there which they are hardly able to control, so the oilseed rape growing seems to have moved somewhat North and West from its original pattern, which was East Anglia, Kent, Southern England and up into Lincolnshire and Yorkshire. Its tending to move further out now to get away from the pests. There will be a number of bee farmers really toiling for forage now that their local farmers aren't growing oilseed rape because of the lack of neonicotinoids. Further out they may be gaining, having never had the luck, if you could call it that, of having oilseed rape before.

Steve: I have seen your spreadsheet records showing the apiaries and so forth. Are you doing weekly inspections?

Murray: Our aspiration is to do them on a roughly 10 day cycle, but the key word is aspiration. A few days of rain, or some people off sick, or a truck breakdown can interfere with it. We try to adopt a system that means that if we are late it isn't critical. We go round the bees on a regular basis, aspiring to do it every 10 days or so, and each round is a full examination with swarm control and disease checks. Give them enough space, do any manipulations that are required.

Steve: Do you record in some way the outcome of those inspections or do you record by exception, in other words only when something is wrong?

Murray: We have an apiary report sheet which the guys have to fill in each time. It includes number of hives at the start, number at the finish, number of splits they made, how many boxes they had to add or take off – it's all on the sheet and is a global figure for the apiary. On an individual hive level, we basically record things by writing on the front of the box. We do a different colour on a different place on the hive each year so that by the time you come back to the same colour and place it's 10 years down the line and the pen has faded off. There are no sophisticated things like hive record cards or anything like that.

Disease and pest management

Steve: So, disease and pest management; what do you do and are there any stories about that?

Murray: Are you looking for hilarious stories?

Steve: [Laughs] Well, there's not going to be many with disease and pest management.

Murray: Oh, I don't know!

Pests and diseases. We've seen most of them now. We encounter on a regular basis nosema of both types, chalkbrood, sac brood, European Foul

Brood, very rarely American Foul Brood, and varroa of course. One thing that's disappeared since varroa has been treated for is braula, which was a parasitic fly that lived on the bees. You saw it on every hive but now we haven't seen it for years.

The biggy of all these is varroa, and varroa brings along with it viruses, or it brings increased susceptibility to viruses that were already there. The big thing for the rest of my lifetime and probably the next generation is going to be the constant war against varroa, and the constant need for fresh and effective treatments.

Steve: Because they get resistant?

Murray: They get resistant to it. There are a lot of fairly interesting and a lot of fairly romantic notions about the breeding of varroa resistant bees. It will happen in time but I'm not holding my breath. Most of the schemes to produce resistant bees at the moment either peter out, or they produce a bee that's of little or no commercial value, because they are so obsessed with hygiene that they don't raise enough brood; they're always picking it out.

There's a long way to go before we have a resistant bee that won't require treatment and will continue to be healthy. A lot of people focus so much on varroa and viruses that they forget that the old killers, especially nosema haven't gone away. They are still there so you must maintain your hives in decent condition and feed them for winter with good sound food that is not going to cause any conditions like dysentery, so you can prevent nosema from becoming a runaway problem.

Steve: That's why you use the invert syrup?

Murray: Invert syrup is ideal. The reasons for using it are several: (1) although it's more expensive than sugar you have no labour involved in making it, it's ready to use straight from the tanker (2) the product never deteriorates because the water content in it is sufficiently low so as not to support microbial growth (3) the bees just take it down and cap it – it's as good as honey and doesn't crystalise in the combs. We'd never go back to sugar unless we could get it for virtually nothing. The lack of microbial growth means that syrup can stay on the hives for protracted periods. It doesn't ferment or upset their stomachs or cause dysentry or anything like that.

[I noticed that in 2018 Murray changed over to using normal sugar syrup made on site, using dry sugar plus water mixed in a giant tank, and asked him about it:

Murray: For very late feeding invert syrup is preferable as it never ferments. That is, in part at least, due to its concentration being

so high that microbial activity does not occur. However, I got a great offer from a British sugar company. The price they offered made it viable to have a mixing system built to turn out syrup at 66% solids where microbial growth is negligible. Yes, it's marginally inferior, but the savings in one year are huge. We saved tens of thousands of pounds in the year, net of the cost of the tank, and with hive numbers increasing by 25% next year the future savings will be greater still.]

Steve: For nosema is it just a case of being a good beekeeper and keeping strong colonies?

Murray: Nosema is a sneaky one. There are two nosemas; the traditional old one Nosema-apis, which used to cause colonies to collapse in winter or fail to build up in the spring, and there's Nosema-ceranae, which is a relative newcomer. It has a completely different pattern in that it gets you in the warm weather, in the summer. They are both around and they both require you to manage bees in a sound manner without stressing them too much, because as soon as they get stressed or start to defecate inside the hive it just becomes a runaway problem. Keep your bees fed with as clean food as possible on relatively clean new combs and then you need to put them on a site where they are not sitting in a cold pool. Sitting in cold dank air makes things worse. These things reduce the risk of nosema but you can't be completely sure that you'll be away from it.

Steve: From all of your years of looking after bees what would you say the worst disease or pest problem you have ever had was?

Murray: On an ongoing basis over the years the thing that has probably lost us more colonies and cost us more income than anything else will be varroa. However, in 2009 this area was hit with a European Foul Brood (EFB) outbreak which was the biggest the UK had ever seen, and we had to destroy 169 colonies that season. It was quite a traumatic event and it affected a lot of people in this area. The truth is, it had probably been around for a while and we hadn't recognised it, because this wasn't a foul brood area and we weren't up to speed with dealing with it.

Since that date that has probably been the biggest cause of our losses because we don't have a huge problem with varroa at the moment. Our treatments are relatively fresh and effective, so in terms of the biggest single incident that caused loss it would be the EFB outbreak which started in 2009 and is still ongoing albeit at a far lower and less damaging level.

Steve: As you were saying earlier when you find it you pretty well "get rid" rather than do anything to treat it.

Murray: There's been many policies about how to deal with foul brood; antibiotics, shook swarms, but our conclusion from several years of experience is that no solution has a sufficiently low recurrence rate for us to consider it viable, given the investment of time and materials to do shook swarming, for example. It's actually far more effective just to cull the stock, sterilise everything and start afresh.

Steve: That year when you had to destroy so many, in 2009, you still did OK that year?

Murray: It wasn't great. The bees were very disturbed, they were constantly being inspected by bee inspectors and they were unsettled, so it wasn't a great year.

Swarm control

Steve: So, what about the thing that you said needs a book all of its own...

Murray: Swarm control!

Steve: Indeed

Murray: Yes, it does need a book all of its own. There are lots of factors affecting how we do swarm control and as I've said to a number of people, we need the maximum number of stable colonies, where swarm control has been dealt with and the colony has settled down, by the first fortnight in July.

You first of all have to see how many winter losses you suffered and then exploit any swarming colonies to restock your empty hives. What we do is move the hive that is preparing to swarm across the apiary to a vacant spot, and then we put a new hive on the position where it previously sat. Then we take 1–3 frames of brood, depending on how advanced preparations are, and the queen away from the swarming hive and put them in the new hive on the old spot. The swarming hive now has most of the brood nest and the queen cells. You can select a queen cell to keep, you can kill them all and introduce a queen, or you can kill them all and insert a queen cell from a line of your choice. That will make up your winter loss as the queen will emerge, mate and start laying.

The safest thing is to knock all of the queen cells down and come back at the next inspection and knock all down bar one after that. It is the safest way but we need to short circuit that and not waste those two weeks so other ways are less perfect but will produce a stronger colony in time for when we need it.

Steve: OK, anything else?

Murray: Well the next way is you lift the brood nest up above a flying board *[note: this can be a spare hive floor, or one of the various boards such as a Snelgrove board, Horsley board or whatever – basically a board to enable a vertical split].* You put the queen and maybe only a frame of eggs in the bottom box, then the queen excluder and supers, then the fly board and your split on top, which you then treat exactly as you would in the earlier scenario. You raise a young queen in the top half and the old queen takes about 3 weeks before swarm preparations resume, so this is a process you have to repeat, but it may be that things settle down and you don't have to. Normally when things are going well with that system you have to do a new split about every 3 weeks.

There is a variation called Demaree which is a very old fashioned one but it works.

Steve: So, in your case you are more likely to let them raise their own queen rather than introduce one?

Murray: We are, but only for one reason – we are not likely to have enough mated laying queens ready to introduce to them all. The majority will end up with, not necessarily a self-raised queen, but we keep the queen cells from our better queens and introduce those to the splits. We also take out cases of queens from Jolanta in the breeding unit and we can take them out into the field and use them. So, saying that we let them raise their own is true occasionally, but it's not our method of choice. The methods of choice are to influence the queen that the colony will be headed by.

Then magically we hope it's got to July and they've decided not to swarm anymore.

[break to make a cup of tea]

The thing is with swarm control, there are so many ways. You can remove nucs, rotate brood nests, there's a lot of ways. For us it's determined by how many winter losses we need to make up, and the date. The date is crucial because you don't want to be splitting too late, because then you end up with two small hives at heather time, and you're trying to avoid that at all costs. The decision can be made on the basis of so many factors, the quality of the bees, and it can sometimes come down to, in the end, what we've got left on the truck. If we've run out of equipment we may have to do something completely out of left field. You just have to have a lot of weapons in your armoury to deal with it. The state of advancement of swarm preparations…there are so many factors that you could write a book on it.

Steve: Yes. You were saying that you then take them to the heather.

Murray: Yes. We keep them under control. We use single brood boxes with queen excluders, and we keep that going and eventually we get to the point where they are coming back to strength again but it's within a couple of weeks of them going to the heather. At that point instead of doing anything remarkable we'll just double the brood nest – give them a second brood box below the queen excluder and let the queen have the run of the place. That tends to stop the swarming, and by the time there's any chance of them coming back to it they are already on the heather, working hard. The heather is hard work for the bees and although you do get heather swarming, it does happen, it's not especially common. It's a pest because you won't get honey off that one, but it's not economic to do anything about it.

On selecting queens

Steve: OK. When we were talking to Jolanta earlier she said that you are the one who selects the queens to breed from, so my question is what do you look for and how do you decide which ones to select?

Murray: A lot of bee breeders have a formal scoring system and they rank certain important properties and give each a score, so that anything scoring below a certain level is never allowed into the breeding process. Our needs are much more basic. We need a honey producing colony and we need the bees to be workable and healthy. One of the properties we look out for is the brood pattern, and fecundity of the queen. You're looking for a solid brood pattern and bees that are steady on the comb and don't run too much, so they are easier to manage, meaning you can run more colonies. You're looking at the honey production, and there are other things like we won't allow a colony with more than a few cells of chalk brood into the process. Ultimately, we would hope to get chalk brood out entirely. We are also trying to choose bees from types that are resistant to EFB because, despite the fact that we are dealing with EFB, and it's not an overwhelming problem for us, it's still there and is definitely linked to bee type. It's not that one bee is immune and one isn't but the degree of susceptibility varies, so we look to breed from stock that has shown itself to be less so.

We buy in breeder queens from other people and we select the best from our own. We keep the good ones from the year before, and from the field where we have 3,000 hives we only need to choose 5 or 6 to go into the breeding program. We can afford to be extremely choosy. "Good enough" is not going to get in; it's got to be something exceptional to get accepted and brought in. They can be rejected on the flimsiest of things, like a couple of bees jumped off and stung me, rather than it being an aggressive hive, but you know, you're going to find one where that doesn't happen.

Steve: It's one of the advantages of being so big isn't it. You showed me earlier on the type of plastic push in cage that you use for queen introduction.

Murray: They are a very good piece of kit.

Steve: Did you do other things before?

Murray: We have experimented but, in the past, most of our introductions were done using the shipping cages. It's pretty successful but the push in cages just give you that 10% extra success that makes it worthwhile. With the value of a home bred queen the cost and inconvenience of using a push in cage is small.

Finding apiary sites

Steve: Right. Next is apiary sites. Finding them. What do you do?

Murray: I've got a strong propensity to go raking about between apiaries, especially if I'm on my own. I tend to gravitate towards larger owners, growers, contract farmers – people who we can get a large number of places from – and then start small and then expand later on once they've got to know us. It's a long-term process to acquire enough good places. At the heather good places are key, and also wintering sites are key. If you take on a poor wintering place you've got just a load of rubbish to work with in the spring, so choice of wintering sites is crucial. The winter sites are the ones I take a lot of time selecting. It needs to get a bit of winter sun, have good air drainage, nice aspect and also a track record. We won't suddenly take a place on the first time and smack 100 hives down. We put a normal sized group in there to begin with, and see how it performs. Then, if it's a good place we will increase the numbers

However, life is full of surprises because some of our best wintering places should theoretically be dreadful. Little boggy patches in the middle of a forest, for example, where people would always expect the bees to die, but they are fine. There are other places with good air drainage which are not very good, but the general rule is: South facing, good air drainage and some winter sun.

Steve: What do you mean by air drainage?

Murray: Cold damp air which forms next to the land, especially in winter, is able to drain away down to the bottom of the valley away from the bees.

Steve: So, it's the dampness that is the problem more than cold?

Murray: Yes. It's a sort of dank penetrating humid cold that comes with it. A long time ago a gamekeeper, who is dead now, was helping me choose a heather site up near Braemar, and he said to me about a site, "I wouldn't put them up there." I was quite young at the time. I asked why not, and he said, "there's the river you see, and the river you don't." The water is the river you can see, but above these rivers there are rivers of cold air that come off the hills and flow down the valleys. I remember putting some hives where he said to go and some others on a beautiful spot that I liked in a flat dry place. I got twice as much honey where he said to go! It was up the hill and it was windy but lower down they just didn't do so well.

Steve: We talked earlier about the temperament of bees and I wondered at what point you decide that they are too feisty?

Murray: If they are attacking your veil and stinging you when the hive next door is not then that's not good. In some conditions they are all unhappy, but because we have so many hives we can pick out the ones that are really unpleasant. You either re-queen them or just mark them down as unsuitable stock to breed from, and let nature take its course with them.

Steve: What about the optimal number of colonies on a site?

Murray: It's a function of the amount of forage there is

Steve: You just learn that over the years?

Murray: Yes. But oilseed rape, for example; you don't want more than about 1 or 1.5 hives per hectare.

Steve: For the heather I guess you just have endless hills of heather, or not?

Murray: Well there are still places where the forage could potentially be limited and you tend to put fewer colonies there. You know where the forage is unlimited; I mean, we've got places where you could literally put 1,000 hives, but we never do, because if that place fails it's an awful lot of hives to fail. So, we generally don't ever have over 100 hives in any one place, and even that is very rare, it's more usually about 40 on the heather sites. Hence there's 70 odd sites and about 3,000 hives on the heather.

Steve: You were saying yesterday that when inspecting hives they do it quickly, something like 6 minutes?

Murray: Yes, the budget for working on the hives including doing any splits if needed is 6 minutes per hive. That's per beekeeper, so if you've got 2 beekeepers on an apiary with 16 hives we expect them to be done in 48 minutes.

Steve: Right. They just know what they are looking for and work quickly.

Murray: Yes. Some of the quickest workers are people like me and Jolanta. The more experienced people get used to it and are quicker, say about 4 to 5 minutes.

Steve: Does that mean just pulling out 2 or 3 combs?

Murray: You actually learn to just scan. You know for example that if you open a hive and get empty comb, then 2 or three frames of eggs, and it's obviously an expanding nest, you're not going to have swarm cells, they're just not going to be there, so you don't have to look much further. You then just need to go to the sealed brood, shake a frame or two and look for any duff cells or discoloured larvae, and then put it back together. So, if it's a problem-free hive it might only take 2 minutes but the next one might have queen cells and require splitting, so that one could take 8 or 10 minutes, so the 6 minutes is an average.

Things that matter or things that are changing

Steve: OK, in general in the world of beekeeping, can you think of things that have changed over the years?

Murray: Varroa, varroa and varroa. There are no other threats around like varroa. People go on about tropilaelaps, which would be unlikely to get established in this country, because after a brood free period of 5 days it's all dead. I don't fear the small hive beetle at all because of the needs for a certain pupation soil temperature, and the Asian hornet – let's see what happens. It could become an issue to the mating unit, but everything else is migratory and during the time that the hornets are at their worst the bees will be on the mountains far away. By the time they come back the season for hornets will be largely over.

The big change is varroa, and although honey prices seem quite high they are historically lower than they were 30 years ago, index linked to commodities. The number of colonies you need to run per person today to be profitable is greater today than it was 30, 40 or 50 years ago.

Steve: Is that due to imports of honey from other countries?

Murray: Not especially. Fashions change and the pressure from supermarket buyers in particular keeps the prices down. If you could make a living off 300 hives in the past, you now need 400. Heather honey has gone up in price significantly in the last couple of years in bulk, but it's still something like 20% lower than it was in the mid 1980s.

Steve: How much does it go for?

Murray: £4 per pound in a steel drum. But if you index link the price from 1984 forward it should be about £5.80, so it has lost ground.

Steve: It's funny isn't it because whenever anybody tastes proper natural honey they can't believe how much better it is than some of the stuff you can buy.

Murray: A lot of the supermarket honey is produced from areas where there is

Wood hives with poly feeders

Grafting shed

Ancient Smith hives

a lot of forage but it is relatively nondescript to a British taste. Someone from, for example Argentina will love Argentine honey, but here we will think it's nothing special. They'll taste ours and think it's not nice but we think it is.

Steve: Can you think of anything that may change over the next 20 years in your profession?

Murray: My feeling is that, once economic austerity relaxes, the positive sentiment towards honey bees will result in possibly some rather bee-friendly government policies. We get very little support at the moment but the need for honey bees is becoming clearer and clearer, and what we can do for the world. Nature's pollinators can't do what we can because populations can't respond as quickly as us moving hives for pollination as required. I hope that policies will make it quite a friendly environment for us. I'm actually optimistic about the future. There are reasons to be pessimistic because of varroa and the lack of weapons to fight it, and the lack of will to provide us with what we need to keep our bees healthy, and increasing levels of regulation – these are all negatives that cause us problems – but I think that there are many people who's ear we have. They are likely to produce policies that are friendly towards us, and we could possibly, in 20 years time, be into a golden era for bee farming.

Steve: Which would be nice for your successors.

Murray: Yes, and nice for me to see too, if I'm still around. So, I'm generally positive about the future.

Working with others...and people I admire now

Steve: Yes. Oh, and do you ever meet up with other beekeepers and compare notes? Is there anyone you particularly admire and why?

Murray: Interesting question. The Bee farmers Association in Scotland are actually quite active. They are not a pressure group but we get together and we talk a lot about how things have been, the harvest, where's been good and where's been bad, and all of the problems that face us. So yes, we do get an idea of how well we are doing, and Nationally they have a spring get together down near Coventry where we have talks and a trade fair and conduct the business of the association, and a lot of information changes hands there. There is a lot of networking that goes on, which is good.

In terms of who do I particularly admire, it's hard to say. There are a number of people who I admire in some of the things they do, but not other things they do. I have actually stolen some peoples' ideas. It would probably surprise this guy to know, but there's a guy in Yorkshire called John Whent.

He is quite a large commercial bee farmer, he is the man who has truly demonstrated the concept of making money from bees without packing your honey in jars and flogging your tail around, because he did it in bulk. The model I went on to in the later years, the last 12–15 years, (the bulk only model) was essentially inspired by realising that he was the one guy who had really made a lot of money from bees, and he was the one guy that didn't pack any jars.

Steve: Right, that's interesting.

Murray: He maintained his total focus on the bees, their welfare, and production and didn't get himself side-tracked with customers and packing and things which in some cases are vanity projects.

Steve: It also means that you are focusing on what you are best at.

Murray: Focus on what you are best at, yes. The calculation that he'd have done in the past (he didn't say he'd done it but it was apparent that he had) was that the time taken on these things was not worth it.

Nobody tells you. They say here's a ton of honey you made and next you jar it, and move it around, and sell it for the highest price you can get. You get as far down the chain as you can, to get as close to the retail customer as you can. What this doesn't take into account is, "okay, how much extra time does it free up and how many extra hives can I run in that time?" The extra value in the bulk volume and the bulk price is greater than taking the smaller volume and doing those extra steps. The extra honey will also be at reduced costs because there are a lot of costs involved with taking the stuff to market. Also, you are not subject to the regulations anything like you are further down the line, with constant interference from health officials and so on. When your honey is in bulk the quality system takes place when it gets to the packer. You have to have a tolerable environment but it doesn't have to be the absolute "white tunnel" environment that you need for packing.

Steve: I suppose you don't see your name on a jar do you?

Murray: No, but that doesn't pay the bills! It's another one of those, it's nice to see, but do you need to see it? Not really. That's more to keep you personally happy rather than financially happy.

Things that rile me

Steve: Yeah. What about, to use an awful pun: do you have a bee in your bonnet about some issue right now, something that really bugs you?

Murray: I've always been, and still am, bugged by this assumption that takes place in beekeeping circles and other places, that the two words "foreign" and "inferior" are interchangeable. This is not just in bees and honey, but also to do with the whole climate around Brexit and things like that, the idea that foreign is bad. When people are like that I do like to point them to a Flanders and Swann episode – a song that they did called "A Song of Patriotic Prejudice" which is absolutely epic. It is guaranteed to offend everybody except the English, and it's brilliant. They have a go at the Scots, the Welsh, the Irish, the Greeks, people who "eat garlic in bed" and, of course, they are making fun of that attitude. I think it's priceless and if anybody has that attitude I point them to the You Tube video.

Steve: OK. If you had a magic wand is there anything you would do to change things?

Murray: Reverse the Brexit result! Probably if I couldn't do that I'd reverse the Scottish independence vote. Actually, that wouldn't be a bad thing at all for me in business because the Scottish government is very "pro bee". They helped us after the hard weather; in 2012 there was a very bad winter and a lot of losses, and the Scottish government helped the professionals in Scotland but the English government declined to do so.

Steve: Surprise, surprise.

Murray: Purely on a selfish level it would probably be beneficial to me. A lot of our product sells because there is a premium over English heather honey, not because it's any better, because it isn't, but because of its provenance. An independent Scotland would be an even higher status of provenance. That's not why I believe in independence though. Since I was about 8 years old I felt I was Scottish, so it's from the heart not from the head.

Steve: Well, you never know, it might happen one day.

Murray: That's that, but in terms of bees and the magic wand, I'd like Jolanta to come up with the perfect disease free, calm tempered queen that guarantees to get a crop in any season when the weather's decent, and doesn't swarm, doesn't cost too much with management problems and if we give it to anybody else the swarms come off and go back to our hives.

Actually, joking apart, we've seen that happen. We've seen our own queens that we've sold turn up in our hives at times. It happened this year in fact. A swarm came into our home base at Coupar Angus with this marked queen in it, it was one of our coloured marks. The next thing somebody came to us and said that the bees we sold them swarmed. I said "well, what did you do with them?" He said he put them in a hive. I asked if he gave them space, and he said that it was just a nucleus. "I told you, in ten days they will need another box." He said,

"I didn't give it one, I didn't think it would need it this year." So it flew away to an empty hive in my yard.

Steve: Yeah, crazy

Murray: I gave him another queen.

Plans for the future

Steve: OK. What about plans for the future? You have expanded a lot, even in the last few years, haven't you?

Murray: The biggest issue facing us at the moment is that we've hopelessly outgrown our premises. They are not in the best of repair but I don't want to spend too much money on them because we're actually planning to move. We've got a new site identified and agreed with the owner. The fact that our existing premises are outgrown and the yard needs a lot of work done on it is something that I'd rather not do if I don't have to. In the medium term we will be building a far larger bee farm. Once that's built it takes the space constraints away, and the scale we can go to. The new place will be 5 times the size that we have now with almost infinite scope for adding extra bays onto the shed, which is not that expensive. From that point on we can start to move forward in whatever way the market leads us, consistent with the family being willing to take it forward themselves.

There are a lot of ways in which we could expand if we want to. We need the space and the capital, so the bees have got to earn a fair bit first, and then we build the new bee farm and take it from there. There's no point just expanding exponentially if there isn't a market for your product. It has to be market led.

Steve: Your family is quite involved then?

Murray: All but one are involved to some extent. My eldest son is in physics and electronics working on radar systems, but he's very switched on and will come over to help with the "business" side of the business. My daughter is very good with the bees and that side of things, and she will take over the management of the bee unit once I feel ready to step aside from that, and of course my second son is very good with mechanics and looks after all the trucks and machinery, so he can carry on with that. The future in that respect is quite secure.

I've got a youngest one who may or may not want to come in. The second youngest daughter is a professional nurse so I don't foresee her coming into the business at all, but the choice is hers if she wants it. I have two grandchildren who my daughter would like to come along after her.

Hives near Newtonmore

Steve: It's going to go on and on

Murray: Hopefully it goes on and on

Steve: What do you do outside of the bee season and where do you go for your holidays? Or is that "not applicable" given what you said earlier about hours?!

Murray: It is a "not applicable!" As I said to you last night I've had 3 holidays in 35 years and they don't add up to 3 weeks between them, and I hated all of them to some extent because they took me away from what I like most. They take me out of my comfort zone, and they make me stressed. I don't really want to go on holiday. It gives me a lot of issues because people would like to go on holiday with me; my children and friends, and they think there's something wrong with them that I don't want to go. It's not that at all, I'm just doing what I love so why would I want to go away from it?

Steve: You are very lucky in many ways

Murray: Very lucky. Normal rules don't apply. I'm doing what I love, I love what I do and if I was offered holiday I'd probably go and look at bees.

Steve: And your favourite honey, you said, was bell heather?

Murray: Bell heather, without a doubt. My favourite honey of all the ones we produce is bell heather. There are some other beautiful honeys produced around the world, not necessarily the UK, and I have tasted some honeys that are absolutely glorious, and some that are utterly disgusting. But, of all the honeys that we do my favourite for flavour and taste and smell is bell heather.

Steve: What was the one I tasted?

Murray: The one you tasted was ling heather.

Steve: That was good.

Murray: I like lime as well. It's another very nice honey that we get. We don't get huge quantities of it, just a little bit, some summers.

Bee keeping philosophy and innovations in design or approach...

Steve: Have you ever invented or designed anything that you use in your business that you couldn't buy anywhere? Actually, I've seen one; your 4 drills in a row for making wire holes in frames.

Murray: Yes! It's not so much that we invent things, but we adapt things. I'll steal good ideas from anybody and we'll put them together in ways that are not quite the same as other people have done. The thing that springs to mind, it's nothing radical, but for moving polystyrene hives long distances you require travel screens. But you also need to stack the hives up on a truck. You need to put them one on top of the other which then blocks off the travel screen, so we've got these weird shaped travel screens that I designed myself. They have a large area for bees to cluster in underneath the screen, and then I've got these weird four pieces of wood on the top, that stick out further than the screen. The two middle ones are lower and the two outside ones are higher but have got notched corners. They are to engage with the floor of the hive above, so that when the floor of a hive above locks onto it the stack doesn't slide. They are locked in position, and you've got air channels through the load to take the air away. The number of people that have taken them away to copy them…so that is one that I designed myself.

Steve: You haven't patented it then?

Murray: Ah no, it's not a patentable idea. Even if it was I wouldn't make money off it so it's not worth the effort. We just let people copy it and get satisfaction from that. A guy in Spain who was selling bees to the north of Europe actually used some of them and successfully moved hives in 40°C, in Spain, and drove them all the way to Finland, and didn't lose any.

Steve: That's amazing isn't it?

Murray: It's quite good, and gratifying that it worked as well as that.

Steve: I think I've asked everything I was hoping to ask

Murray: Oh, it's amazing that we got through your list so quickly.

Steve: I've got one slightly corny one, which you may not be able to answer, but is there something in a nutshell that you could say about your approach to bee keeping? Is there a particular philosophy or "the Murray McGregor way" of doing bee keeping? What's it all about?

Murray: It's a difficult question but when it comes down to a philosophy about bee keeping, it's that the best way to keep bees is the way that the bees would like to be kept. An awful lot of people are guilty of anthropomorphic thinking about bees – they have their own belief structures so therefore they will keep their bees that way, be it to do with organic or natural or whatever. Some people do homeopathic bee keeping…The thing that will tell you how bees like to be kept is the bees themselves. We think that we need rest, we need holidays, it's good to have quiet times, but bees don't like that. Bees like to be working. When there's nectar around and the weather is fine the bees go at it hammer and tongs all day.

You have to recognise that they are not being forced to work. They work because they want to. They breed when they want to, they swarm when they want to and you have to work with the bees' system and not against it. As soon as you are fighting your bees you are not getting the best out of them. You are always trying to exploit the bees' natural urges to do what they want to do, and go with the flow of the bees, and channel them in the way you want rather than regard bee keeping as an exercise in mastering control. That's just not the way to get the best out of them.

Steve: No

Murray: It's the same with people. A lot of people say that I am very soft with my staff but I get a lot out of them. I see good aspects in them and I see weaknesses, and I prefer to focus on what people can do well for me rather than obsess about their faults, and try to treat them well. All these people who have come to work for me, young kids from Eastern Europe and things like that; the way I treat them is to think how I would want my kids to be treated if they were away from their country. I get enormous reward from it, and they never leave!

Steve: Is there anything else you'd like to say?

Murray: Well it's almost impossible to get who we are and what we do into a short piece, so be sure to email me for any clarifications you need or perhaps I can explain things in a more clear way. You've seen a lot today, and you have probably realised that what looks like a relatively simple business is actually quite complex.

Steve: It's been a privilege, for sure. Thank you very much.

Murray: Remember, you'll not get any bullshit, or lies, and no subject is off limits.

Twenty three tonnes of heather honey in barrels.

Steve: Apart from perhaps that estate up the road?

Murray: Oh no, the fact that we are on that estate is not off limits. It's no secret. Anything that goes on with the family, behind the gates, that is off limits, but the fact that we go there is not. It's been on the Duchy Originals honey jar label for 25 years so they've been quite happy for it to be common knowledge. The status that sometimes gets assigned to me, the Royal Beekeeper, was not my invention. They are my bees on the Royal land; I'm not keeping the Royal bees. I never was.

That issue with the antibiotics

Steve: The worst thing about the media is that if I google your name the first thing I'll get is a page of stuff about the Royal Beekeeper being taken to court or something.

Murray: Oh, that was the antibiotics thing.

Steve: It's incredible that of all the things that could come up it's always that – it's just the way it works.

Murray: The interesting thing about that was that some newspapers printed the whole story but some did not, and just printed alarmist stuff. The truth of that was that a European foulbrood outbreak came along and we'd had a big meeting with officials, and it was discussed that we were going to treat the bees with oxytetracycline to stall the outbreak until we could get it properly dealt with. There was no schedule given; we were supposed to be getting the supply from the Government, and I was on such a scale and so dependent on things that I couldn't wait, I had to deal with it. They finally delivered it sometime in October; we needed it in August!

I bought a private supply. I secured exactly the same stuff privately, prepared it in line with best practice. What they don't tell you is that I made the contacts with America, I had done the research for them, I had prepared a document with all the application protocols and the way it should be done, collected from best practise in North America, and I taught the Scottish Government official in charge of it how to do it. I then had to attend a training course where he trained me! I did all that, I sourced the product for them, I found a supply and gave them all of the details, and they had a paperwork problem. They had to go somewhere else. I used the approved medication and they ended up using a non-approved medication!

If it had been left to the Scottish authorities this would never have happened, but somebody down in England said, "Oh, he's using oxytetracycline which he got without a prescription – sue him!"

Steve: Right.

Murray: That's what it came down to. I fought them a bit and got it toned down a lot but...the Press report afterwards didn't report what the judgement was, they reported the original charges but actually most of that had been chucked out. I got a moderate fine for buying it and possessing it, and I got no fine at all for using it. The judge said in the court, "this is a red tape offence." To read the press reports you wouldn't think so – you'd think I'd been selling toxic honey or something like that. Nobody cared about the judgement at the end. He said he had to fine me because I technically was guilty and I pleaded guilty, but he said it was a red tape offence.

Steve: How long did that go on for

Murray: Four and a half, five years.

Steve: Oh dear.

Murray: More than that. The final verdict was in 2017, and it was a 2009 event. They asked me about my use of oxytetracycline and I told them that it was both a government supply and a private supply, and I produced the documents and showed them the invoice from America, so I gave them the information myself. But they actually ended up turning me into more of a folk hero figure locally, rather than a villain. I now get so many calls giving me information and asking for advice, it's embarrassing.

Steve: If it took so long it must have been stressful?

Murray: It became stressful just a few months before, when the court papers arrived, but even then, I just got my lawyer on the job. I produced lots of evidence and was able to prove that some of the stuff written by the people preparing the charges was actually outright false. It didn't stop the press regurgitating it, because it was on the original list of charges submitted by the court. A lot of it was absolute nonsense.

I was able to get enough to treat every hive in the outbreak for $320. The Scottish Government's treatment cost £20,000. Outrageous. Such a waste of taxpayers' money.

Then they tried to get me for a non-approved varroa treatment. That got kicked out quite quickly because it was given to me by a bee inspector, as a gift! He was a bee inspector in England. He came up to Scotland in his DEFRA van to do some private shooting, and he must have had 5,000 strips of this stuff in the back of his van, so when they found out that led to a quick dropping of charges.

They should have dropped the lot because it didn't do them any good in the end, and despite these searches, it's done me no harm whatsoever.

Steve: No. Well, that's because any reasonable person whose livelihood depended on bees wouldn't have had any choice; you either let your bees die or deal with it.

Murray: European foulbrood was spreading like wildfire and we needed to get it under control before we could really deal with it. We were getting into a time of the year where it wasn't symptomatic anymore, because the brood was finishing off, so you couldn't diagnose, but there was still drifting taking place between hives, so it was still spreading around. What we did was to arrest the spread.

Steve: Then the next year you had to destroy a lot?

Murray: Well it wasn't nearly so many because we culled the biggest number that year. The number that showed up the next year was not huge. It's stayed relatively low ever since. I think this year we had 38, but compared to 169...and it's 38 out of a higher number of hives. If it was the same number of hives we'd be looking at about 18. We have got more and more hives since then, and will continue to have more.

Steve: OK

Murray: Is that you done?

Steve: Yeah, thanks very much.

Afterword

It was a privilege spending time with Murray McGregor. I also gleaned a few other nuggets of information not covered by the interview which I am happy to share.

Murray uses polystyrene feeders on the top of his hives, even the wooden ones, because they provide excellent insulation over the winter. These are the feeders that cover the whole area of the top of the hive and can hold a lot of sugar syrup. His hives come back from the heather late in the season and have to be treated for varroa and fed. What Murray does, and I have not seen this anywhere else, is to use this late season winter feeding as an opportunity to get new frames of foundation drawn. This is how he described it:

Murray: You can get three frames of foundation drawn dead easily in autumn, even five. Take out the combs you wish to replace and shake any bees off and set aside. In the MIDDLE of the colony insert the frames of new foundation, never two together, so they are interspersed with brood frames like (O for old, N for new) OOONONONOOO, assuming eleven frames. Feed the bees a full winter feed and check what has happened to the foundation a few days later. In September it is normally perfectly drawn and there will be lots of eggs and young larvae in the centre of the combs. In October it might have less or no brood.
The big advantage of autumn comb replacement is that they are always perfect. There is no drone rearing at that time of the year so the combs are drawn 100% worker. You won't find that in any text book!

Another practice that I found interesting was what Murray does when uniting two colonies. He does not use the newspaper method but instead sprays the top of one box and the bottom of the other box with air freshener, then puts the two together. Unless there is a good reason he leaves both queens alive, allowing nature to take its course. He said that when uniting a split (in which a daughter queen was raised) back to the original hive, by the spring they find the young queen in charge 80% of the time, the old queen 15% of the time and both 5% of the time. Having two queens in a hive is much more common than many beekeepers realise.

Finally, many beekeepers in the UK today are taught to replace comb in their hives very frequently because apparently dark comb is a terrible thing. In fact, this is not often the case. One practice that is gaining in popularity is that of carrying out a "shook swarm" on a colony in spring in order to get all of the combs changed at once. The professional beekeepers that I have spoken to about this, including Murray, say that this is very definitely not something that they would do in their hives. At the time when you want your colonies to grow rapidly through spring in order to be strong for the honey flows the idea of shaking all of the bees onto foundation frames and destroying their brood is anathema to the people that I have interviewed. Shook swarms are sometimes used when European Foul Brood is found in the hive, although Murray, who has lots of experience of that disease, prefers to destroy the colony. In his experience EFB keeps cropping up even if you do carry out shook swarms.

Strong colonies

Michael Palmer, Vermont, USA

'Once I started raising my own queens I would say that the change was noticeable immediately. I raised queens in the summer and next year those colonies were amazing'

Introduction to Mike and re-queening nucs

I stayed with Mike and Lesley Palmer for a few days in July 2017. My daughter Clíona, who is my beekeeping assistant when she is not being a teacher, was happy to tag along for my trans-Atlantic adventure, and I was grateful to have her company. Long distance travel is not my favourite thing, particularly the endless waiting around at airports, so I always try to bring somebody with me. We flew from Manchester (UK) to Atlanta and then on to Burlington in Vermont.

Atlanta has a huge airport. The soft Southern voice over the tannoy was matched by the friendly smiles and the Georgia accents of the shop assistants and waiters. "I love your accent, are you British?" Clíona was asked as she purchased a rather fetching garment with "Atlanta" emblazoned across the front.

We managed to resist the huge bags of Reese's peanut butter cups, tempting though they were, and headed to a Chinese restaurant called P.F.Changs for a delicious chicken fried rice meal. The service was friendly, helpful and prompt, and the food was great. It felt right to be in America.

I had never met Mike Palmer but as we came through the arrivals lounge at Burlington he was easy to spot with his ponytail and spectacles, which I recognised from pictures and videos on the internet. He had not only agreed to be interviewed in the middle of the busy beekeeping season, but he drove for an hour to meet us at Burlington airport on a Sunday night and insisted that we stayed with him at his family home. I warned him that he might regret that decision because there are rumours that I snore. Loudly, apparently. He replied, "I said you can stay at my house, not sleep with me!"

Mike's house is at French Hill Apiaries on the edge of St Albans in a beautiful forested area. There are cedar tiles on the sides of the house with a slate tiled roof and a red brick chimney pointing skywards from the centre. We were welcomed by a large white wolf-like creature which turned out to be his pet Maremma sheepdog, called Wilson. It took a few days for Wilson to get used to us. I soon realised that the couch was strictly for use by dogs, not humans. There were cats too, who were most displeased at being moved to the floor from their cosy chairs.

One thing I noticed very quickly, as an avid tea drinker, was the absence of an electric kettle. They did have tea but clearly coffee is the drink of choice in the Palmer home and there is usually a jug of hot coffee on hand. There was a very large mason jar of Mike's honey to the side of the coffee machine which I quickly sampled. It was good honey as expected; proper honey from proper beekeepers is always good. I placed the tea bags, which I had brought with me from England, as is my custom, next to the honey. Home is where the teabags are.

Sleep came easily when I was in Vermont. Mike, in common with all farmers, is an early riser and, weather permitting, he likes to be out working by 6:30am. I tried to keep up but I am a night owl so could usually only manage about 8am. It was two or three days before I actually sat down to formerly interview Mike, because a combination of jet lag and disorientation meant that I did not trust myself to ask questions coherently. As it turned out, like every other beekeeper I have met, Mike could talk for hours about bees with very little prompting from me.

Upon waking on my first full day at Mike's, I staggered downstairs to munch on muesli and gulp down tea. I was amazed to see about fifty queen cages lined up on the coffee table, each emitting loud buzzing sounds. These were queens with attendant workers that were waiting to head colonies.

As the days passed the number of buzzing boxes decreased, until on the day I departed for home I think just a couple remained.

That first morning Mike drove us to his cell builder apiary in his enormous RAM 3500 pick-up truck. Everybody in the area seems to have a pick-up truck and Mike's is a particularly fine example. The top of the tyres are at waist level. The apiary is next to a corn field hidden amongst trees. There are a lot of trees in Vermont. Cell builder colonies towered seven or eight boxes high on one side of the site, and the other was peppered with nucleus "brood factory" colonies made up of brightly coloured four frame boxes. The nucs were five boxes tall and in groups of two; two colonies side by side with a shared roof and entrances pointing opposite ways. They thrive in this configuration, each colony's warmth being shared with its neighbour.

The mission for the day was re-queening the nucs. Actually, my mission was to take photographs and ask questions. Mike raises his own queens, nice big fat beauties which are generally quite dark in colour. At this time of year, he will carry out regicide in the woods of Vermont, but only when necessary. Any queen that is not performing will be killed and replaced. He checks the amount of brood and the brood pattern as part of the process of determining whether Her Majesty will be granted a stay of execution.

Mike uses shorthand notes written on duct tape stuck on each hive's roof. It's yet another use of duct tape in beekeeping (there are so many). I soon got the hang of his notes and was able to read the history of the colony. "WDQ" is short for "White Dot Queen" which in this case meant a queen raised in 2015, so the note,"KWDQ" meant "Killed White Dot Queen". Generally, the notes show whether brood or honey frames were added or removed, for example "-2FB" (minus two frames of brood), when new boxes or frames of foundation were added, changes to the queen and, rarely, comments on temper. A comment is not needed for good temper!

Where the bees have decided to re-queen themselves, either by supersedure or swarming, the queen will be a "NDQ" (No Dot Queen). As long as they are performing well Mike is happy to leave these alone. After all, a thriving colony that re-queens itself is about as good as it gets. If he is not happy with her she will be for the chop; it's just that finding her can be a challenge. In such cases out comes the shaker box; a brood box with a queen excluder nailed to the bottom with duct tape around the inside at the top, and the bees will be shaken through it. I helped with this but at the time my eyes were being uncooperative as I had something called posterior capsule opacification, so my queen spotting was not up to par. My daughter fared better in this respect, and even managed to spot an elusive virgin, which was impressive.

Even when I know that a queen needs to be changed, for whatever reason, I still find it difficult to squash her. Some of them look so perfect. However, looks can deceive, and the only way to keep colonies bursting with vitality is to maintain the quality of the queens. There is no place for sentimentality in farming, and make no mistake, for Mike Palmer, this is farming. I have to admit something here; when I have a colony of bad-tempered bees that ping my head and sting my ankles, then I find killing the queen an easier task. Thankfully such occasions are rare. With his incredible number of nucleus colonies, Mike Palmer has good quality spare queens on hand as and when needed.

I should say something about the technique used by Mike and his team when shaking bees through a shaker box, which happens a lot in late summer, because tiny details can make a big difference to the outcome. Mike is adamant that frames should be kept in the correct order, meaning the order they were in upon opening the hive. Unless there is a specific reason for muddling things up he puts frames back into the hive in the same order as he found them, thus causing less disruption for the bees.

The order of play is as follows;

- Firstly, a wooden board is placed on the ground to the side of the hive.
- An empty spare brood box goes on the board
- On top of this goes the shaker box, with the queen excluder being the division between the two boxes.
- The hive to be shaken is dismantled, with all except the bottom box being removed and put onto its upturned roof
- Frames are taken from the bottom box and shaken into the shaker box.
- The shaken frames go into the spare brood box beneath the queen excluder, in order.
- After shaking the frame Mike slides the shaker box across creating a space to insert the shaken frame into the box below.
- The frames are pushed along so that there is a space for the next frame, then the shaker box is slid back into place
- The advantage of this is that frames are kept in the right order and any brood on the shaken frames will attract bees down through the excluder.
- The older bees will be flying and the younger ones will gravitate to the brood below
- Bees are persuaded with smoke to move down to the box underneath so that the queen can be found.
- She will normally be moving around a bottom edge, where the excluder and side of the shaker box meet.

- Mike has duct tape along the inside of the shaker box at the top which he says dissuades bees from climbing out over the sides
- Once the spare brood box below is full of frames it is returned to the hive and the empty box takes its place.

Bees being shaken in this way send out great clouds of Nasonov scent which I find very pleasant. Also, this is where the temper of the bees will become most obvious. Assuming the weather is fine most colonies can be shaken with few stings but occasionally there will be a vicious one, making it even more important to find and replace the queen.

This brings me to the topic of queen introduction. I don't fully understand why, but when a colony of bees loses its queen one would expect it to gratefully accept any new queen provided, for without one they are in trouble. In the past when I have introduced new queens into a queenless colony the bees have killed the imposter on occasion, perhaps as often as 30% of the time. This is never going to be acceptable in a commercial honey farm. I was therefore very keen to learn how Michael Palmer introduces queens. He uses push-in cages which he makes from wire mesh and cutters kept on the back of his truck.

"You want to find a frame with emerging brood on it and ideally some nectar," Mike explained. The new queen is taken from its cage, which Mike does out in the open using his fingers, gently holding the queen by her wings or thorax. She is placed on her own under the wire mesh cage, which is pushed into the wax of the comb so that she is trapped there. The emerging brood will naturally recognise her as their queen and she will be well looked after and come into lay over the next few days. The frame with the new queen is placed back into the queenless colony and is marked on the top bar with a marker pen so that it can easily be found. As the hive is put back together the box containing the new queen is also marked on the outside with the word "Here" and the date. Four days later the frame with the new queen on it can quickly be found and the queen released, subject to a quick check for any eggs outside the push in cage, which would indicate the presence of another queen in the colony.

The push in cage method of queen introduction works very well for Mike. I tried it when re-queening my apiary and I got 100% acceptance. In my case I often find that the queen is no longer under the push in cage four days later; she is somewhere in the brood nest laying away. The workers must dig underneath the cage to release her. Mike thinks this could be because I placed the cage too near to the edge of the frame. One of the benefits to me of writing this book and meeting experienced commercial beekeepers is that I have learned so much.

I doubt that I will ever be a large scale beekeeper but I can learn from those like Mike, who handle more bees in a week than I do in a year.

Over the next few days I spent time in Mike's apiaries and enjoyed chatting to his team of helpers. I met Kate, Zac, Rob and Jeremy and was impressed by their friendly camaraderie and beekeeping skill. They are probably used to visits from strangers; apparently Mike has plenty of them now that his talks at the National Honey Show have become so popular on the internet. It is clear that Mike's workers hold him in high regard. They are all experienced beekeepers, some having over 80 colonies of their own.. I noticed that they occasionally wander over to Mike to show him something, a frame of brood perhaps, to get his advice on what he wants them to do. I think that is the sign of a good working relationship. There is respect on both sides and a shared goal, which is the prosperity of the bees.

On one sunny afternoon we drove out to Swanton near Lake Champlain to visit a place called Maple City Candy. Mike insisted that I try a maple creemee, which is a delicious maple syrup flavoured ice cream in a cone, although apparently in Vermont it's bad form to call them anything other than creemees. The maple syrup industry is very big in Vermont, and as I found out in my interview with Mike, it's what he worked on before becoming a commercial beekeeper. It turns out, unsurprisingly, that honeybee colonies are much more interesting and rewarding than great vats of boiling maple sap.

Just before I get to the interview, I must mention some of the interesting features at Mike's place. Firstly, upstairs he has an amazing collection of bee books, some very old, collected over the years and lovingly referred to in conversation quite frequently. Mike loves Manley's books especially; a common trait of the UK honey farmers I have met.

Downstairs Mike has a grand collection of honey pots gathered on his travels all over the world. In his garden hang several bird feeders and we were treated to the sight of some lovely birds not found at home in the UK. I saw humming birds, which are tiny, and a beautiful crimson cardinal. The other less attractive but ever-present flyers were mosquitoes. Luckily, they don't bother me too much and as a veteran of holidays on the Kyles of Bute in Scotland, where the vicious midge is king, I hardly noticed them.

Talking with Mike

When I finally sat down to interview Mike I was sprawled out on the hardwood floor by the coffee table that was home to his caged queen bees, and he sat back on his sofa. The recording device was appropriately placed. The occasional buzz from the queen cages or a bark from Wilson were the only sounds picked up by the microphone other than Mike's steady clear voice as he answered my questions.

Good years and bad years

Steve: Have you got any memory of any particularly amazing or terrible years? You've been beekeeping a long time.

Michael: I have, of course

Steve: Do any stick in your mind?

Michael: 1986

Steve: Was that good or bad?

Michael: Bad. Eighty sixed in '86. Did you ever hear that phrase, eighty sixed?

Steve: No

Michael: Well, it was pretty bad

Steve: Really?

Michael: Yes. That's the year I bought Chazy Orchards' bees. I bought 500 colonies of bees, for over $70,000, and we had such a poor year. It rained all summer and we made only two and a half tons of honey. That's only a few pounds per colony. I had to get a job that winter; I worked in a factory making syrup utensils. I never want to see another 1986 again.

Steve: Was it the weather, is that what did it?

Michael: Yeah, of course, the weather, combined with the way I was keeping bees then. Back then we were splitting all the strong hives in the spring to make nucs, and buying queens from Georgia and Texas and Florida and California to put in the new splits. That's how I was maintaining my numbers. So we made the splits and then it rained all summer. The poor bees in the splits couldn't build back up again, so even if we had had a good flow I don't think we would have made much, because they were struggling to get built back up all summer.

Then comes along this year, and we haven't had a great year – it was snowing in May and we have had over 19 inches or rain for the season...

Steve: Which year is this?

Michael: This year [2017]. But we don't split colonies in the spring anymore, we try to keep them as strong as possible, so when a flow comes along they've got the population to gather it. If you split them in spring they don't get the population to be able to store honey. I think that's probably the biggest difference.

Steve: Any amazingly good years that spring to mind?

Michael: Yeah, I remember the year I made 58 tons, I think it was 2005. Everything hit that year – every flow, they gathered; we had enough rain, it was warm enough. It was something like 135lbs average across all those colonies. That was amazing. I'll probably never see that again.

Steve: What is average?

Michael: I like to make about 100lbs per colony, then I feel like I've got a full crop.

Steve: So is that three supers?

Michael: Yeah. Some will be five and some will be two.

Steve: I think I read somewhere that the smaller colony is the one that you use to make nucs?

Michael: Correct

Before beekeeping

Steve: I think I probably know the answer to this, but what did you do before you started beekeeping?

Michael: I made maple syrup, that's one of the things I did. I graduated from high school in 1967 and then I went to the University of Vermont and enrolled in the pre-veterinarian course. I completed the first year pretty well but in the second year I just...I wasn't into it anymore.

Steve: Was it because it was theoretical rather than practical?

Michael: I never did any veterinary work. It was all the beginning couple of years of college, where you take English and Math and political science...

Steve: Not really what you wanted

Michael: Not really what I wanted. I'm a hands-on kind of person. That was about the time of the "back to the land" movement. People were coming out of the cities and suburbs, moving to the country and trying to live in the country.

Steve: So you did maple syrup?

Michael: Yeah, I started sugaring in 1971, working for people, and by about 1974 or 1975 I rented a side of a mountain that was covered with maple trees and was all set up to gather sap. Actually, I had another sugar wood before that with my wife and another couple. We rented woods and had an evaporator, we had a tractor with chains and a wagon with a big tank, we shared it together. But after that I went out on my own. It's hard to have partners, yeah? [smiling]

I gathered sap and made syrup for a number of years. I took it down to New York City and that area, and sold it to health food stores. Because I didn't own the property I had to pay rent, I had to pay for fuel and help, so by the time I did that there wasn't a whole lot left.

By that time I had a couple of bee hives...

Steve: Just for fun?

Michael: Yeah, just for fun, and that seemed pretty interesting. I think I got bees in 1974. So by about '76 or '77 I'm gathering sap all day, boiling it up all night, watching those little bubbles and breathing that sweet steam. Your body gets full of sugar – absolutely wired on sugar, and you go home at 2am but you can't sleep because of it. I thought, "here it is at 2am and I've been watching those little bubbles, and I've got bees at home, and they collect their own nectar and process and pack it, what am I doing here?"

So I decided to get out of that and get into bees bigger. That was the beginning of trying to be commercial. I think that was in 1977, the year my Mom died, that was the last year I boiled sap.

I got a government loan to buy another 100 hives, so by 1980 I probably had a couple of hundred. A job opening came up in '82 over on the west side of Lake Champlain at Chazy Orchards, which is billed as the largest McIntosh apple orchard in the world. They had 500 colonies to manage, and I got a job managing their colonies. In the previous five years they had gone from about 850 colonies down to 150, because of American foul brood. They had to burn everything and then rebuild. They had a beekeeper that did that for them but he didn't want to continue, so I got the job.

When I took over they were making no honey, and after four years I had them back to making a honey crop again. They were paying me $110 a day in 1982 to '84, and that's a pretty good wage, but I was only working forty days a year. I could surely do better than that if I had my own operation. Our first daughter was born in '87, and I wanted to be able to contribute to the family income. It's a good thing my wife was a registered nurse because she could pay the bills when I couldn't.

Steve: Then you bought the colonies didn't you?

Michael: Certainly did. In 1986 I told them that I would buy their bees and rent them back for pollination, so they still get them for pollination but they would be my bees, and I'd get the honey crop. The owner of the orchard wasn't too excited about that but his son talked him into it. They had a lot of problems getting and keeping a qualified beekeeper. Part of it was the old man; he couldn't see the wisdom of paying a decent salary for a beekeeper, so he put the lowest guy on the totem pole in charge of the bees, who knew nothing about bees. That's why they went from 850 to 150.

So I said I was either going to buy their bees or I'd be gone, and they realised that it would be a better plan to go along with me. That was in '86; after pollination in '86 they became mine. That was at a time when interest rates were really high, they might have been 18%, and farmers around here were really struggling to get operating cash. People were buying farms and losing them because they couldn't pay the bills.

The state of Vermont had a plan to re-finance agriculture loans. I applied for and got the re-finance loan. Originally, I was going to have to pollinate for seven years, at $22 a hive. I used the loan to pay the orchard, and I can still remember the monthly loan repayment. It was $907.62 every month. Then as the prime rate fell, loan rates fell, but I elected to keep my payment the same $907.62. Eventually it got all the way down to 3.5% interest, and I still paid $907.62, so I paid off the loan real quick. The loan officers just couldn't believe it. I was the only one of all the agricultural loans they had that left the payment the same as it originally was. I paid it all off in 3 or 4 years.

Steve: Great

Michael: Yeah, that was a real help.

A couple of years ago our lieutenant governor, who was obviously going to become governor one day, would go out and do common jobs. He'd go to a gas station and change people's tyres for half a day...

Steve: So that people would vote for him?

Michael: Exactly. He's a good guy. He's not from the party I usually vote for but I would vote for him because I think he's honest. Anyway, I invited him to come and work bees with us for half a day. It was amazing, he rolled up his sleeves, we were making up mating nucs, and he dove right in with no fear. It was very impressive.

Afterwards he said, "Mike, what can we do for you?" meaning what can Vermont do for you. First off, they had done so much already. Vermont food products are worldwide and renowned for quality. Plus the state gave me that re-financing loan and saved my business. I never would have been able to pay for that, with such high interest and having to pollinate almost for free for seven years, I don't know if I would have made it. So, I said that there was really nothing he could do for me. "You've done everything that I could possibly ask for."

Then I thought, perhaps they could re-establish a proper bee inspection program, a honey bee inspection program. As the number of novice beekeepers increases the threat of foul brood and other diseases increases. It took three or four years before we got it, but we now have a reasonably good inspection program with a chief inspector and two seasonal inspectors, which is more than we had in twenty years.

That guy was actually elected governor last November.

Steve: So you're on first name terms

Michael: I actually am. We say hi to each other now. That's what is so special about Vermont. It's so small that you can get to know anybody who's anybody.

Steve: Would you say that the bee inspection program in Vermont is now ahead of other states?

Michael: No, probably not ahead of other states. In some states the beekeeping industry is way more important than it is here. So, they put more resources into it, maybe several seasonal or area inspectors, say a chief plus ten or twelve area inspectors, which is really what we need.

Mentors

Steve: When you first got your couple of hives, when you were doing maple syrup, did you have a mentor or did you read books? How did you know what to do?

Michael: Well, first off I took a short course. The university extension service offered a course, and at that time Vermont had an extension beekeeper, supplied by the university. That position has now been done away with, which is part of the problem now with beekeeper education – it's so lacking.

What did you just ask me?

Steve: About your first couple of hives, for fun.

Michael: Oh yes. So, I took a course, first it was in the federal building where the post office is, where all the floors are granite, a typical government building, then they moved it to the city elementary school. Can you imagine? An elementary school with young children, and all the floors are carpeted? That seems kind of odd. We were sugaring at the time and we'd go right from the woods to the class. We were working in mud and snow. I'd have mud up to my knees and would leave tracks on the carpet. So, we stopped going.

After that it was asking people that were beekeepers for advice. Around that time there were a number of us starting at the same time with the same intentions, and we would compare notes, and if an idea worked well we would all adopt it.

Steve: So it wasn't any particular book or anything like that?

Michael: No, it was mostly hands on learning. I wish I had read some books, I wish I had had a mentor, a proper mentor or a parent or an uncle that was a beekeeper before me. Mostly it was just trial and error, the hard knocks; you make a mistake, and sometimes you have to make it two or three times before you realise it was a mistake.

Steve: Do you mentor others now?

Michael: Yes. For instance, you came here. We have a constant stream of people coming from all over the world really, mostly from the States, but from all over.

Steve: Was it your You Tube videos that started that?

Michael: I would say it was the You Tube videos of the National Honey show that started it

Steve: Well you've started something there – you're not going to be able to stop them coming now

Michael: That was my intention. I did not realise that fame would follow me around, so I stopped answering my phone and emails. I get hundreds of emails a week, I just can't keep up with it.

Steve: It's interesting isn't it, that video travels around the world in seconds, it's so powerful. Books and articles in magazines are much slower.

Michael: Yeah, as you say it does go all over the world. People write to me saying people are keeping nucleus colonies, and they mention my name – people who speak different languages but I can see it's my system they are using, and I'm blown away by that. I didn't realise that would happen, but I'm glad I did it. I'm glad they put those videos online, I think it's been a real big help. I think in some way I have changed the dynamics of beekeeping. What more could I ask for?

On going commercial

Steve: Here's a bit of a strange one. You know those people who get all enthusiastic at the start, and decide they want to do bee farming for a living – I know these people exist…

Michael: Absolutely

Steve: I'm sure they find out that it isn't as easy as it looks

Michael: They start out with a few hives and do pretty well, and then maybe ten hives and they make a nice honey crop, and they figure, "well I can have a hundred or five hundred." They invest everything into one hundred or five hundred hives, and then they realise that they don't have the experience to run that many. Five is not five hundred. It's a totally different life.

People ask me about this [going commercial] all the time, and I tell them no, don't. Be patient. "Patience Grasshopper." Take a step back, go and find somebody to work for. Go work for a commercial beekeeper for a summer if you really want to do this, and find out what it's really about, how many hours you really have to put in. And see the failures. You need to see the failures as well as the successes. You can have a bad year. You can have two bad years in a row and then you lose it.

Steve: Yeah, you're gone then.

Michael: You can make a bunch of money one year and spend it all, but what are you going to do next year if it's bad? I have a guy who helps us once in a while. He's a border patrol agent, and he's sick of his job – he's out in the woods in winter time on snow shoes, patrolling the border with Canada. People come across with money, drugs and guns.

Steve: OK

Michael: And he's tired of it.

Steve: Well, yes, I'd be tired of that after 5 minutes

Michael: He has it in his head to get into bees; get a hundred hives. He showed me his business plan, and he's going to have so many hives, so many queens and make so much honey. I said, "Where's your truck?". "Oh yeah, I forgot a truck." "Where's your processing building, your honey extractor? Where are you accounting for a bad year or two? Right now, you are working for the federal government on an amazing wage, amazing health insurance, and you have a four-year-old daughter and an eight-year-old boy."

Steve: Yeah

Michael: I said, "if you give up your day job I think you're crazy." You need to work into it slowly, and build up to a hundred in your spare time. Once you get to that, and you sell enough nucleus colonies and make a honey crop, then maybe think about going commercial. He can retire at, I forget, fifty years old or something, which is in about ten years. So, get yourself built up to a hundred hives in the next ten years and then think about it.

I think he listened to me because he stopped talking about quitting his job and being a commercial beekeeper. He got some hives and it was going well and he was on top of the world, then they had a bad year. And he started to see: "Oh my God, everything's swarming, what am I going to do?" and there you go.

I get that a lot.

Steve: It's a business rather than a hobby isn't it?

Michael: You can manage ten or fifty hives and keep your day job, easily on the weekends, but when you have five hundred you can't do it on the weekend.

Steve: No.

When you bought up the colonies in the orchard, presumably you had "done the math" as they say – you'd worked out the numbers – how much honey per hive, the costs and so forth?

Michael: Sort of. It was a struggle at first anyway, going from say 200 to 800.

Steve: So did you have to have help straight away?

Michael: I had help straight away.

Steve: That's a whole different thing isn't it? Managing people as well as managing bees...

Michael: Right. When I had that bad year, I forget how much I owed my helper, but he's a funny guy, he likes to be paid at the end of the year.

Steve: At the end of the year?

Michael: Yeah. He's in a landscaping business, and he would get everybody's bill at the end of the year. But I couldn't pay him at the end of the year, so I gave him the interest that he would have made if he'd invested it, and he was OK with that. Some people wouldn't accept that. Thankfully he could.

Steve: How many hours a week would you say you work?

Michael: All of them!

Steve: What does that mean?

Michael: That means I get up early and I work 'til late.

Steve: Seven days a week?

Michael: Seven days a week.

Steve: You have to love it to do that.

Michael: I rarely get a day off. I take a day off when I'm made to. My daughter got married and I had to take a day off! [laughs]

Operations and revenue streams

Steve: What about the split of your revenue? Is it mostly honey, queens & nucs?

Michael: Yes. When we make a honey crop the revenue from that is huge, and much bigger than the rest. For instance in the last two years I averaged 30 tons. That's a huge amount of money.

Steve: Yes

Michael: But you don't make a big honey crop every year. I forget exactly what it is, but I used to have to get a 40 pound crop per production colony to pay my bills. That was before I sold nucs & queens. Now I sell enough nucs & queens to pay my labour bill, which is huge, it's for sure the largest single cost I have. So, having my help covered, whether I make a honey crop or not, has really changed things.

Steve: Great

Michael: Now I actually have money left at the end of the year. Sometimes a lot of money. Last year I bought a new truck, to lower my tax burden and I needed a new truck; my old truck was sixteen years old. It's really changed everything. Diversification of your income is really important. If you look at any farming, say dairy, it's the same thing. Dairy farmers selling to the packer, or bulk purchaser – they're getting a terrible price, in some years they make less than it costs to produce. But if they are able to change that milk into a finished product, a value-added product, they're going to be ahead, and it helps to insulate against those bad years.

Steve: Of the nucs and queens, it seems to me that queens would be more profitable, because it's one insect, a small package in the post, is that right?

Michael: But the amount of time involved in raising those queens, from cell builder set ups and grafting is huge. I spend every day from the first week in May until the first week in August raising queens, running cell builders and mating queens – it's a huge amount of labour.

Making nucleus colonies, provided I have the queen…I can make up 50 nucs in a day, and put the queens in. Then ten days later I have to check on the queen, see that she's laying, and replace her if not. The amount of actual hands on time with making nucleus colonies is pretty small.

Steve: Whereas the queens are a big deal

Michael: Exactly. So, if I sell 175 nucleus colonies for $200, which is $35,000 and I sell 800 queens like I did last year, at $30 each, it's less money [$24,000] but more time. But they fit together because I need to raise queens to make nucleus colonies, so all I'm doing is raising extra queens to sell, and I always have spare queens on hand.

Steve: Who buys the nucs, is it mostly beginners?

Michael: I would say so, beginners, back yard beekeepers, part timers... sometimes bee clubs – they might buy 20 at a time. I did at one time sell 100 nucs at a time to a company for a few years, until I realised how much they were charging on the resale. They were making more money than me, and I was doing all the work, so I stopped.

Steve: That reminds me of something I was going to ask you, about pricing. Some people are more willing than others to raise prices and you seem a bit reticent to, I don't know, push it.

Michael: I think my prices have traditionally been a bit low.

Steve: Considering that your product is probably a bit better than most...

Michael: I decided this year to raise my prices up to what other people are getting, and they didn't even blink. I'm hoping that price transfers into my production hives when I get ready to sell them. If a nucleus colony goes for $200 should not a production colony go for $300? I would think so, with all the ancillary equipment, and supers and so on. Can I get that? I don't know.

Steve: I hope you're not planning on that anytime soon

Michael: No, not soon.

Steve: One day.

Michael: Yeah. It really depends on how much help I can get over the next few years, whether I cut back a little bit. One of my helpers is thinking about taking a school teaching job and the other is talking about going back to school to get a masters degree...if I have to go back to the drug addicts again...I don't really want to go back to that again.

Steve: No

Michael: These last few years have been so nice, with good help, and I don't want to go back. So, if it comes to that, I can sell off my New York State bees. Some of them are an hour and a half away. I have 450 hives or so over there. I could sell those and focus on my Vermont bees, but still do my nucleus colonies and my queen bees.

Steve: In a way that's a little pension fund sat over there in NYS.

Michael: Yes it is. That's why I'd like to get as much as I can. We don't get a pension.

Steve: Do you ever compare notes with beekeepers in other countries to see how it differs?

Michael: Yes. Our friends in New Zealand, they are just flabbergasted that we don't get a pension. I get no government pension after working as a farmer for my whole life. The only thing I have is the sale of my bees. I hope I don't stay in it one year too long, and have a big catastrophe or something right when I'm getting ready to get out of it.

Steve: Yeah

Michael: Like a big varroa crash, or a new disease comes along. We don't have that great an inspection program, so what would happen if a beginner started keeping bees in used equipment by my breeding station?

Steve: It could completely ruin you

Michael: If they got foul brood into my mating nucs...I don't know if I'd ever rebuild it again. So yeah, I'm concerned about that quite a lot.

Steve: Can you get insurance for that?

Michael: No.

I don't use antibiotics and haven't for twenty years. I used to use them, Americans are famous for using them; I used to dust the inside of the hive to keep foul brood away. I haven't done that for years and years, and I haven't seen any foul brood in my hives.

Steve: How many hives and apiaries do you have?

Michael: I don't know exactly. I have somewhere around 750 production colonies and at the height of the season, oh my God, I must have 1,000 nucleus colonies. That gets slimmed down after we catch the last queens because a lot of the mating nucs get united down for winter. So my 650 mating nucs might drop down to 325.

Natural beekeeping

Steve: When I look online at certain beekeeping forums it seems that some topics are very popular, and there's one that I think is called "natural beekeeping"? Have you got any thoughts on that?

Michael: I don't really know what natural beekeeping is, it's just a term, a catch phrase which gets people to look at the book or whatever. The only natural beekeeping is if they are in a tree and you don't do anything. I don't know. What would natural sheep farming be or natural apple farming or anything else?

I don't know. Let nature take its course? I'm not really in favour of that because I think we're managers of our bees. To let your bees die because you want nature to take its course, and some day you'll be blessed with the perfect bee, because nature has taken its course – I don't agree with that. I think that as a manager of a farm you do the best job you can, and when problems arise you do your best to put them right. That doesn't necessarily mean dosing them with a chemical, there may be other ways, but sometimes that's the only way.

Steve: I agree.

Michael: I have a dog. You give your dog a rabies shot, you feed your dog. So, with natural beekeeping does that mean you don't feed them sugar? What happens in a drought year? There's no honey. If you feed them sugar does that mean it's not natural beekeeping? I don't know. But you feed your dog, you don't just leave it to eat lemmings and mice. You give your kid medication if they get sick. What's wrong with giving your bees some medication when they are sick?

Do I want to rely on medication for my problems? No, of course I don't. Do I think breeding is the way to solve some of the issues we have now in beekeeping? Yes, I do. I and others are working on it, but it doesn't seem to be there quite yet.

Some things are pretty easy to control, for instance brood diseases.

Steve: Chalk brood?

Michael: Chalk brood, even foul brood. Very hygienic bees will clean up so fast that it doesn't become contagious. So, if you have highly hygienic bees you'll never see brood diseases, that's easy. Tracheal mite was a problem, but by breeding from the best colonies in the spring it's now just a minor issue. As happened in the UK.

Steve: The Isle of Wight disease?

Michael: Yes, Acarapis woodi, the tracheal mite. It wiped out loads of colonies but within 5 years it became a minor issue. Varroa mites, oh my gosh, how many people are working on breeding a bee right now that tolerates varroa mites? Lots of people, lots smarter than I, and where is it? Nowhere yet.

So, in the meantime, you don't let your bees die, you do what you need to do to keep them alive whilst working on a breeding program, or others work on a breeding program – some with a lot more backing than I have – so no, with natural beekeeping I don't want to get stuck in that catch phrase. I'll go with it when I can go with it, but I'll offend the natural beekeepers if I have to.

Steve: Well, I guess you probably have a few more bees than they have

Michael: Here in Vermont, and in fact in the whole North East, we have a number of beekeepers who strongly believe in the natural beekeeping dogma.

The last time we had a severe drought the natural beekeepers lost all their bees. In a good year when the bees can build a large population of young bees for winter, the bees can outbreed the varroa mites, but when you have a year like last year, when you have a drought and the bees can't raise a sufficient amount of brood...all the mites jump into that small amount of brood and it's game over.

Steve: Right

Michael: So, it's really dependent on conditions. When you see bad conditions, you need to do something about it, you don't just let all your bees die, I'm sorry, I can't agree with that. If it was true that if we let all the susceptible bees die and we were left with resistant bees, it would be like tracheal mites – it would be a five year program and we'd be rid of this. No. It hasn't happened yet. It's been 30 years since varroa mites invaded North America and where are we now? We're at the same place we were 30 years ago.

Insulation and ventilation

Steve: OK. How about another topic people on forums get excited about, which is insulation and ventilation? A lot of people in the UK think that upper entrances are a strange thing, some people say you should use poly hives and some say wood...there is a difference between the American and UK literature on this. American literature goes on about ventilation whereas in the UK it's more about insulation, and in fact people laugh about the old practice of putting matchsticks under the crown board to provide ventilation. So, in the UK it's more about insulation and in the US it's ventilation – what do you think about that?

Michael: I don't really know what to think about that because our conditions are so different. We have this continental jet stream which comes down the Hudson Bay and freezes us big time in the winter, so we have that and you don't. You have maybe a damper climate, so I don't know what to think. Personally, in my experience I don't think you need to insulate the body of the hive. Our bees are going through long winters, sometimes four months with no cleansing flight, temperatures in the 20 to 30°F below zero, and the only insulation I have is on the crown board so that respiration moisture won't condense and drip down on them.

Other than that, bees don't heat the inside of their hive, they heat their cluster. If you put thermocouples, heat sensors, in the hive, you can see that once you get a few inches away from the cluster it's almost ambient temperatures. So, I wonder how much hive insulation is really going to help. We've always felt that if you insulate the hive, and you finally get a nice day, it takes longer for the cluster to heat up and be warm enough for them to be able to take a cleansing flight. That's how it's always been presented to us. Some beekeepers in the US have double walled hives with sawdust inside, and the things sweat and don't do well.

Steve: OK

Michael: I mean look at some of those old-time beekeepers who have had so much success – they have the rattiest old rottenest bee hives, full of holes – you'd think that the damn bees would just perish, but they don't, so I wonder. If you looked in the old literature, like say "the Hive and the Honeybee" or American Bee Journal…somewhere there is a piece by Farrar – he was a professor in somewhere like Wisconsin – and he wanted to show that insulation isn't necessary. So, he took bee boxes and cut out the side panels, so the bees were totally exposed, except they had a roof, and he put a mesh around the body so that predators would not get in and eat the bees. These things went through the winter with temperatures many degrees below zero – they were just heating the cluster.

Steve: Wow

Michael: So, do bees need insulation? I don't think so. What they need is a way to vent away the moisture that they are giving off all the time. In this climate, if you don't have an upper entrance to get rid of extra winter moisture, the inside of the hive is just soaking wet and mouldy. They need a top and a bottom entrance to get a flow of air. If you want moist air to leave out of the top you need it to be replaced with dry air, from the bottom. I don't favour entrance blocks in the winter. I leave them wide open, with a hardware mesh with about a half inch gauge, so the bees can go through but not the mice.

Steve: OK

Michael: My hives stay absolutely dry.

Steve: You have solid floors though?

Michael: Solid floors. There's so much moisture coming out of that top entrance, sometimes on a cold morning you can see a horizontal icicle sticking out of that entrance, so much moisture is leaving and it freezes as it comes out. That moisture could be trapped inside the hive.

Steve: Yeah.

Michael: But, you have an Atlantic climate and we have a Continental climate, so it's just different; it's hard to compare. When I'm on the UK forum and somebody says something, I have to keep quiet, unless they say something specific about how or what we do here in the US – then I correct them if I can.

Steve: Sure

Michael: But I can't just say "you're wrong" because you have a different climate to us.

Neonicotinoids

Steve: OK. There's one other slightly controversial matter, neonicotinoids. We have a lot of oilseed rape in the UK. Many commercial beekeepers put their hives near it and say there is no problem, but then there are people who study this and say that "the bees are dying"...do you have neonicotinoids here?

Michael: Of course. We have a lot of corn here [maize] and most of that is treated with clothianidin which is a neonicotinoid. We have soya bean here – almost all of it is treated with imidacloprid, which is another neonicotinoid. My bees are surrounded by corn and soya beans, although not 100%, I have plenty of forage out there for the bees. You have seen my bees. My bees are pretty healthy; they are strong and productive. I think the problem at least in my area is not the neonicotinoids but the loss of forage caused by farmers turning meadows and hay fields into corn and soya bean fields, so the pasture is being eliminated. If you get right down in the valleys all that is left is hedgerows. There's no more pastures anymore here. Cattle stays in the barn 12 months per year, they don't need the pastures anymore, so they dug them up, put drainage in and grew corn and soya beans.

If I'm making a hundred-pound crop and I have neonicotinoid crops near me, I can't believe that those neonicotinoid crops are damaging my bees. If I'm making 100lbs and making a 2–6% winter loss, I really don't think so.

Steve: No

Michael: I too know people who have hives on oilseed rape, making 100–200lbs honey crops. I could say, well the neonicotinoids don't affect my bees because they don't forage on corn, and in this climate and soil type they don't work on soya beans, but those bees are actually gathering nectar from rape plants that are treated with neonicotinoids, and I can't believe it's affecting the colony when they are making 100–200lbs honey crops. I just can't believe it.

I mean, yes, it's an insecticide, and insecticides kill insects. You can feed it to them in labs and kill insects. What's happening in the field? Nothing, as far as I can tell. It's too bad that now everybody is focused on neonicotinoids, and they're not thinking about forage. I think that's the problem, our loss of forage.

Steve: OK

Michael: I got appointed by the governor to be on the pollinator protection committee. We have to come up with a plan to protect pollinators, both managed and unmanaged, honeybees and bumble bees and all sorts of other insects

Steve: When did that happen?

Michael: In the fall of 2016 through to spring 2017, to come up with some guidelines. Almost the whole thing was about neonicitinoids – how to regulate, should we ban them? I think we finally came down to agree that we don't want to ban them because they are an important tool for the farmers...

Steve: Yes

Michael: But we want to regulate them. Rather than every cornfield in Vermont being planted with neonicitinoid corn, you have to show need, so you have to do a survey of the field that's going to be planted with corn, to find out the wire worm load, and if it reaches a threshold number then you can order your corn with neonicotinoid in it. I would rather see that than see farmers spraying organophosphates all over the countryside and killing every flying thing within range. At least it's targeted

Steve: That makes sense

Michael: Whether the legislature will adopt our suggestions we'll have to wait and see. But the seed companies make it very difficult to obtain non-treated seed. So, if we require non-treated seed, will that mean those seed companies stop doing business in Vermont? I don't know, we'll have to wait and see. It wouldn't surprise me; they're making money on this. But we have to do something about the prophylactic use of neonicotinoid treated seeds. Fruit orchards sample to see if they need to treat, beekeepers sample to see if they need to treat, I'm sure dairy farmers look for mites and things on their animals to know if they need to treat. They don't just treat all their cattle.

I was on the fence at first, about this. My bees are healthy, so neonicotinoids are a non-issue for me, but now it's getting into the water – they're finding it in the local streams at 1.5–1.7 parts per billion. It's very low, but it shouldn't be there. I think one report said that at 1 part per billion you start to see changes in the biology of life in the water. If that's the case I think we really need to do something about prophylactic use. Let's not ban it but use it just when necessary.

Steve: On the forage question, do you think there are things that can be done to help? For example, in the UK some local authorities are planting wild flowers on edges of roads and parks – I'm not sure if that would help honey bees though?

Michael: I guess that would depend on what you planted. It should be a mix of things that help all of the pollinators. They're planting small leaf lindens in parks and cities, which I think is a great thing.

Beekeeping in cities

Steve: That's reminded me, do you have much beekeeping in cities here, on rooftops and so on?

Michael: Not in the North East because we have so much land, but if you go down into the cities like Boston and New York, where it is so crowded and therefore not a good idea to start an apiary in the small back yard you have, there are bees on rooftops. Some cities previously outlawed keeping bees but now with rooftop apiaries they have been re-approved. They don't bother anybody up there. I know people who have rooftop bees in New York City and they say they make 100lbs of honey, which I find hard to believe, but they say they can. They have Central Park which is a huge park.

Queen breeding

Steve: OK, so I'm jumping around a bit but...when you are deciding which queen to breed from, what are the main characteristics that you look for?

Michael: I like to tell people that they should pick the five most important characteristics to them. I've seen some lists that are so long I don't see how you could possibly breed for all that, so I like to keep it to the things you can control. That would be things like temper, diseases present, honey made (combined with how much sugar you had to feed) – if you got 100lbs of honey but then had to feed them 50lbs of syrup the actual production is 50lbs not 100lbs. What else? Possibly if you were working on a varroa program it would be the change in population of varroa over time. Some colonies might go from 2 mites to twenty to a hundred; others might be something like 2 mites to three mites to four mites, on different samples, and while they have mites the population isn't exploding – you could keep track of that.

Steve: Yeah

Michael: I like to keep track of the queen – what queen line it is, when introduced, if they changed her, if they formed swarm cells. I think you can select for colonies that have a lower propensity to swarm. Not that they won't swarm; every colony will swarm at some point, we will never have non-swarming bees.

Steve: But for a commercial beekeeper swarming is pretty bad, right?

Michael: I think that swarming is probably the greatest reducer of overall honey crop.

Steve: OK

Michael: So, we keep a yard sheet. We keep track of these things. Every colony is numbered, and those factors are recorded onto the yard sheet. I can go through the yard sheets over winter and figure out who to breed from next year.

Steve: Is that something you do on the computer at home?

Michael: No, I'm rather computer challenged, I find it just as easy to go through each yard sheet. The first thing I look at is honey production.

Steve: Yep

Michael: For each yard I want to find the top honey producers. Then I look at how much sugar they got fed. I look at the net honey crop, how they did over winter, that I see no chalk brood in there and they don't staple my socks to my ankles when I open them up – those kind of things. It's all on the yard sheet.

If you think about it, every single colony is potentially a breeder, and you look for reasons to take it off the list. If it has an "X" under temper, it's off the list. Or if it has three X's, or a skull and crossbones! If you have that colony you work that one last. Sometimes I get to the yard and see a hive with a skull and crossbones on it and a note saying "this queen must die!" [laughs]

Steve: You don't have too many of them?

Michael: Not too many.

Steve: When you first bought those bees from the orchard, the 500 colonies, were you doing that then – selecting queens?

Michael: No, I was not breeding queens at that time, I was buying them.

Steve: But when you did start doing this [breeding your own queens] how long would you say it took to go from a big variability of colony performance to being consistently good?

Michael: Once I started raising my own queens I would say that the change was noticeable immediately

Steve: Literally?

Michael: Literally. I raised the queens in the summer and next year those colonies were amazing.

Steve: Right

Michael: That was the beginning, back in the late '90s. Maybe it took about 5 years, to around 2003, to get a consistent stock. I am not an annual or semi-annual re-queener of beehives. I believe that smart bees know how to re-queen themselves.

Steve: OK

Michael: A smart beehive re-queens itself when it needs to, and maintains its production; that's the kind of bee I want. So I let those bees maintain themselves, they're all out there, plus my additions...pretty soon things are going well. They are not all identical, they are all different, but their overall production, quality and temperament, they're all pretty good. So I would say 5 years. You start a queen rearing programme, along with a nucleus colony programme – it's difficult if all you do is raise queens and run production colonies – but with nucleus colonies you can have so many queens, and so many nucleus colonies – you have the numbers, and numbers matter

Steve: Because you know what "good" looks like?

Michael: Of course. Eventually you have so many different lines, some going on for years superseding themselves, that you can select from

Steve: And you're not able to find an isolated mating place?

Michael: No, I tried going over to New York State where there aren't any other bees, but it would take me an hour and a half to two hours to drive over there, and the same back home, it just didn't seem to make sense. There's two theories on that one; either go to an isolated place or flood the area with your own bees. I'm in favour of flooding the area with my own bees.

Steve: Yes

Michael: If I go to an isolated place, what am I supposed to do? I've got to take a number of mating nucs, of whatever it is I'm trying to get mated, and a number of drone mother colonies, both of which I have to select. But what about the colonies left behind? There's got to be something of value in there. Even though they may not be top producers in their apiary they may have something that will be lacking if I restrict my breeding, my genetic diversity down to say 10 mother drone colonies and a few hundred mating nucs. I would be really limiting my genetic diversity.

Steve: OK. So, what you're saying is that you are essentially a farmer, just because you are a beekeeper doesn't mean you are not a farmer, and your bees are your livestock, and the quality of that livestock is fundamental to your success.

Michael: Right

Steve: And one of the biggest threats to your success is, say, a hobby beekeeper bringing in other stock that's not of such quality and it mates with your queens – is that correct?

Michael: Yes, but I don't know if it's so much the hobby beekeepers. The hobby beekeepers only have a few hives and if they are close to my breeding station I can offer them queens to re-queen their colony or two colonies or whatever, that's not so much the problem. I think the larger problem is commercial beekeepers – larger migratory operations, that will move in 25 or 50 colonies near to my mating station. This has happened, and it's just rubbish stock from down South. They winter their bees down South, so they have Italian type bees which are hugely prolific, they have huge colonies of bees from which they make lots of nucleus colonies to sell to our local beekeepers.

They can buy queen cells from Florida and don't have to raise their own stock, or they can buy cheap queens from down South and not raise their own stock. Then they can go back South in the winter time and re-adjust everything, get everything built back up again and then come back up here next year. That's more of an issue to me than a few hobby beekeepers.

Steve: Sure

Hobby beekeepers

Michael: The thing I worry about with hobby beekeepers is not genetics, it's foul brood. They don't know what it is, they don't look at their bees often enough, they are suckers for used equipment – that scares me more than anything I think.

Steve: And are you seeing some of that?

Michael: I haven't seen foul brood in 20 years, but my operation is just hanging there, waiting. I don't use antibiotics and it's just a matter of time before I get an infection. If it's in a production yard I can control it, even if I had to burn the whole production yard it's a loss but not a huge loss, I can re-establish 25 more hives, but to get it in my mating nuc operation, that would be huge.

Steve: Yes

Michael: I don't know if I'd want to rebuild those 650 mating nucs

Steve: Someone would

Michael: I think that might just retire me

Steve: This is the thing isn't it, being in business, you are vulnerable to so many things – the weather and all sorts.

Michael: It's just the way bees are. If you are a cattle breeder the insemination guy comes along and you can breed for all kinds of things, and you get pure breeding, but bees aren't like that. You can put your cattle or sheep on your field and you can be pretty sure they're not going to get any disease from your neighbours flock or herd, but that's not the way with bees. Bees fly so far; the whole neighbourhood within 2 miles of the apiary is their territory, and they can pick up something from anywhere in there.

Steve: Yeah

Michael: I mean, I'm running a breeding operation here, and if that commercial guy with Italian stock from down South plops his bees within 2 miles of me, within mating distance of me – it makes it really difficult.

Try explaining this to a dairy farmer. They are smart and know their business, but they don't know about bees, they don't even think about bees. So, when you say to them, you know, try running your cattle breeding program if bulls had wings, then they get it. You start talking to them about nucleus colonies and they look at you funny, "what's a nucleus colony?", try explaining that. Ah, they're heifers – then they get it. So, you just have to tell it to them in their own terms, then they get it.

But that doesn't make it any easier

Steve: No

Michael: I could use instrumental insemination, I guess, but I don't really want to get into that. I'd just as soon do it naturally. Getting back to your natural beekeeping [laughs].

Steve: OK, something I was talking to you about earlier: when inspecting a colony, with experience, you were saying that when something is wrong it stands out quickly. How do you get to that point?

Michael: It's just experience, looking at thousands of bee hives for years. It's no different to working in a factory where it's so loud you can hardly hear the person next to you, and there's so many noises in the factory – all the whirs and whizzes and dings and clinks – but all of a sudden there's a noise that's not supposed to be there, and you actually hear that noise, and it alerts you that something is wrong. All those noises that are right, you don't really hear them anymore, but you hear the one that's wrong.

It's the same with bees. When you first start you're looking at all the things that are right; are there eggs, is there brood present, those kinds of things, you don't see the things that are wrong. You have a checklist in your mind, and you have to go through this whole entire checklist, of things that are right, but once you've looked at enough bees you train yourself to see the things that are wrong, and ignore the things that are right. So, you're looking at the comb and you see an egg; you're not thinking, "are there eggs?", you just see it, but that capping, that brood capping that has a hole in it – that should focus your attention.

Steve: Yeah, what's going on there?

Michael: You open it up to see what's inside. Is it a white pearly larva? Yes, then it's healthy. Is it off coloured, is it twisted, is it chalk brood...or you start to see other holes in other capping's...a lot of times that hole is just because they haven't finished capping the cell yet. If it's chalk brood you'll see other cells with chalk brood, you just have to look.

Steve: Yeah

Michael: It just takes time. It takes years of looking at bee hives. Same thing as working in a factory; I used to work in a factory making maple syrup cans, and that's where I got that experience of hearing things that are wrong in an environment where you can't hear yourself think.

Steve: OK, I'm sorry to talk about things that we have already talked about [off tape], but how about swarm prevention? I know it's not entirely possible, but we've talked about breeding less swarmy bees...are there particular weather conditions that make them swarm more, or are there swarmy years and not so swarmy years?

Michael: I would say being confined in rainy weather is conducive to swarming.

Steve: That's at particular times of the year?

Michael: Yes, in the swarm season. We're having rain now [late July] but we're not getting swarming. But in May, when we had three or four days of rain then one nice day, followed by three or four days of rain, then one nice day...

Steve: Basically the UK climate!

Michael: Swarming was terrible. But hopefully it got done early enough so that they can build up and make a crop on the later flows. There's really nothing else you can do. You manage them the best you can, and you do the same thing to every hive, and most of them will respond, but a number won't. If you have time you go back and adjust it again, but sometimes there isn't time to do that.

Steve: You were saying yesterday that you give them extra space quite early, is that right?

Michael: Right. You know, there has been so much written on the causes of swarming, and I don't know if anybody really knows, not the whole story. It just seems to me after all these years that part of the story, maybe the main part, is that when nectar is being stored in the active brood rearing cluster, in the cells where brood is being reared, and if they can't move it out because there is nowhere for it to go, then swarming happens. That can be in the spring before supers are on, or in the summer on a good flow and the supers are full.

What happens is that they bring in the nectar, they put it anywhere they can, and they are frantic to get back out to get the next load of nectar. So, it goes into the cells where brood has recently emerged, but later on they move it up out of the way, and start to ripen it, and brood rearing continues. But if they can't move it up and out of the brood nest then swarm preparations start.

When I first started I was taught not to put supers on until after the first dandelion

Steve: Right

Michael: After the dandelion we've already had a couple of good flows; we've had maple usually, and other trees that flower in the spring, and then dandelion comes along and we can get 30–40lbs of honey from dandelion. So imagine a double deep; the brood nest is located in the top deep in the spring, those flows are coming in, and the only place to put it is in and around that brood nest. By putting 2 supers on early, during the tree bloom, we are providing them a place to put nectar coming in from spring flows.

When we didn't super until after the dandelion we'd reverse the brood chambers at the end of April before moving them to the apple orchards. We'd count the number of frames of brood – not by counting individual frames but by looking up from the bottom and finding the first frame with brood on and the last, and count all those frames between to come up with a number, and any colony with 8 or more was a possible split. We would take a split away from that. We used to do that in the orchard, and then after the apples we would take them back to the apiary and reverse them again, because by now they are getting crowded. And now you've got dandelion nectar coming in and the only place to put it is the brood chamber which triggers swarming.

So now we get supers on early. I don't pollinate orchards anymore, so I don't need double deeps – I can have double deeps with 2 supers on top because they're not moving. So we put the supers on early to get the tree bloom and the dandelion bloom, and that acts like a reversing, so the bees can move up onto empty comb. If the queen feels she wants to move up and lay in the supers, so what? I don't use excluders.

Steve: No?

Michael: I think queen excluders select for the least prolific queens because prolific queens will swarm with an excluder unless you do something to that colony like split it or take brood away. So they move up and by the end of the dandelion you have honey at the top in the supers, then the brood cluster and then empty space in the bottom box, so now it's time to reverse the brood boxes to put the empty space above the brood. At that time, if they've been working well, we can put another one or two supers on. Once the bees are making honey and they have space to store it, it seems to take care of the swarming impulse, and they just get on with making honey.

Steve: Do you arrange combs in the supers so that there is empty comb in the bottom super next to the brood box?

Michael: It depends how far along they are in filling their supers. OK, so we've got the 2 supers on, and we haven't reversed yet. Now we reverse, and take the colony apart. If the top super is filling well with honey and the bottom super has some brood in it, maybe in the middle 3 or 4 combs, if I leave that on the brood nest the way it is that's just going to continue. Now if they're working well they are going to need another super; if they're really working well in the supers and if they are plugging up every cell in the hive with nectar then I'm more than one super behind, and I'll need to add two more.

What we'll do is reverse the two supers because the bottom one has some brood in it and the top one is all nectar and honey. We'll put the super with all the honey on top of the broodnest, then the super with some brood in it, then on top of that a new super with empty comb in it. If they had queen cells started, or if they are putting burr comb between boxes, which means we're late, then we'll put 2 supers on – one under and one over the existing two. So now they have the new one, then the one with lots of honey in it, then the one with brood in it, then another empty one on top. You want to keep them always moving up, always moving up.

Steve: Yep.

Michael: Yes, I cut queen cells out, but it depends on the age of the queen cells and what condition they are in. Cells with eggs? No problem. If they are not sealed most colonies won't swarm yet so I'll cut them out, and continue to provide space with supers and reversals. As long as they have space above and good weather and a good honey flow that takes care of it. If we have days and weeks of rain then all bets are off.

Steve: So then what?

Michael: Well, if there are sealed queen cells then you have to look deeper. You have to see if there is a queen still in that hive. The chances are they may have swarmed. You look at the bee population and you look at how much brood is in the hive – could that population have raised that amount of brood and put that amount of honey in the supers? Maybe not.

I showed you yesterday that frame of sealed brood where there was a patch in the middle of 3 or 4 inches of emerging brood, where they are just beginning to emerge. That's where the queen loves to lay eggs the most, so you go for those places and look for eggs. The colony could have swarmed yesterday and its still got eggs, but if you look at a comb where the brood has recently emerged, and it has eggs, and you were there all morning so you know they didn't swarm today, then you have a queen in there. Then you can go ahead and cut sealed queen cells out. I think that's the easiest way, rather than trying to find the queen.

Steve: Yeah

Michael: If you've only got a few hives then go looking for the queen, and find her. If you have hundreds or thousands of colonies you don't have time to look for the queen, so you have to look for other ways to assure yourself that there is a queen in the hive.

Introducing queens

Steve: OK, with introduction of queens, sometimes it's into a small nuc and sometimes it's a massive colony. Have you learnt from mistakes over the years about the best way to introduce queens?

Michael: Introducing queens is not always easy. You just have to think about it. I mean, why wouldn't they accept a queen? You've taken a laying queen out of the hive, and you are going to introduce a new one, but what type of queen? One like these in cages [points to 40 queens in cages with attendant workers lined up on coffee table]? I took a laying queen and put her in a cage and left her on my table. The next day she's all shrunk up. She won't smell like a laying queen, she's not in lay anymore. So, you take a laying queen out of a colony and put one of these new ones in, in a cage with some candy, and expect them to accept her even though she is a non-laying queen. That's an issue.

The only thing you can do is somehow get that queen into lay. You can make up a nuc – we make up hundreds every year – and we make up the nuc and take them to the apiary and put a queen in using a cage with candy. Because they are weak, and they know they're queenless (they know within an hour or two), they'll accept the queen. But a big production colony, they have so many old bees in there, they won't accept a queen. They'll kill her every time and try to raise their own.

Steve: Yeah

Michael: To introduce a new non-laying queen to a big production colony that had a laying queen…you have to get her into lay first. So, you can introduce her into a nucleus colony and after she has been laying for 2–3 weeks you can use that nucleus colony to requeen the hive. That works very well. But you can't always make a nucleus colony.

Steve: If you were going to do that though, would you just lift the frames from the nuc into a box in the hive?

Michael: Yes, you can remove some non brood combs from the top brood box and put in the frames from the nuc, and that works very well, because she's a laying queen.

Steve: No newspaper or anything?

Michael: No, it's not needed, but if it makes you feel better go ahead and use newspaper. Sometimes it helps – putting a nuc in doesn't always work, sometimes it fails.

I find that the easiest way to re-queen a big production colony is not to make a nucleus colony, because you have to remove frames, and what do you do with them? You have to put them in storage and they get eaten by wax moths.

I use a push in cage. I find a frame with emerging brood on it and maybe some nectar, and I put the queen on it under the push in cage, which I push into the comb to the midrib, so she's trapped in there on that frame with emerging brood. Then I mark the frame and put it back into the colony, and in four days I come back and pull the comb out. I look outside the cage for eggs, because sometimes there's two queens in a hive. So you look outside the cage for eggs – if there are you have to find the other queen – if not you just pull off the cage, and she just walks calmly around and lays eggs and is immediately accepted.

Steve: Nice

Michael: One way to do it if you only have a small number of hives is to take off the supers, then take off the top brood box, then shake all the bees from the top brood box into the hive, so you just have the frames in the top box with brood, pollen and so forth, but no bees. It will be to the side on an upturned roof or something. Then you put the supers back on then a queen excluder on top, then put on the brood box with brood but no bees, and close it up until the next day. Overnight the nurse bees will have moved up through the queen excluder to get to the brood, and of course the queen can't get up there.

The next day you take off the top box, remove the queen excluder, and put on a solid crown board (cover up the hole if there is one). We use a notched cover board which makes an entrance, but you don't use those, so what you want is to put back the top box onto the solid crown board and have an entrance pointing the opposite way to the main hive entrance [note: could use a spare floor instead of a notched crown board]. Now you can put a queen into that top box in a queen cage with candy, because any older bees will fly out and return to the hive below, so what's left is just young bees, and young bees accept queens...almost always. So that queen gets released and starts her broodnest, and in three weeks the brood has emerged and it's getting strong. Then you go down into the hive below, kill that old queen, and unite the two colonies using newspaper. There was no nuc to make, no frames to remove, and you had two queens laying for a while which helps to boost the population. You also get a second chance, because sometimes the new queen turns out to be worse than the one you want to replace, so you get a chance to judge her when she's in the top box. When you unite them together it works very well. I like to do that – I did it for a while, but now I've got so many bees that it's really hard to do that, so we use the push in cage, and get almost 100% acceptance. As long as you look for eggs outside that cage!

Steve: And the beauty of selecting the frame that you do, for the push in cage, is that it is exactly the type of area that any outside queen would want to lay in.

Michael: Yes, but you can see if they have accepted her anyway, by looking at the bees on the cage. If you can just move them off easily with your fingers then she's accepted, but if they are gripping tight onto the cage then they haven't accepted her. You pull them off by their wings and its like they're stuck on with velcro – then they haven't accepted her. And the reason must be there's another queen in the hive – it happens. I've seen 30%, by shaking bees though excluders I've seen 30% of colonies with multiple queens. That's huge. That would mean 30% of your re-queening would fail, just doing direct re-queening.

Steve: In my personal case it probably does [laughs]

Michael: I think it happens a lot, and people don't realise. They stop at the first queen. They look for the queen to replace, find and remove her, and then stop. They don't consider that there might be another queen in there. I've even seen three – I've seen two daughters and a mother queen in a hive before. It happens every year, that we find eggs outside the cage.

On making mistakes and meeting challenges

Steve: OK, so to give hope to those of us who aren't as experienced as you, have you got any tales of woe where you did something stupid, or something that went wrong to prove you're not perfect!?

Michael: I am far from it [laughs].

Yeah, I've gone back to an apiary after it rained overnight and found a colony where I forgot to put the crown board and lid back on [laughs]

Steve: So that happens

Michael: Yeah, that happens, really stupid stuff happens. Or I forget what day I'm on, so I forget to put a comb in the breeder hive, so when I go to graft I haven't got any grafting material. I don't know, I've made every mistake. That's why having a mentor, or a relative who got me into it, would have been helpful. I usually make every mistake at least once.

Steve: One thing, it's not really a mistake, but yesterday you showed me a frame of queen cells that had been destroyed by a virgin in the cell builder hive…was that one that flew in from outside?

Michael: I think so, that happens

Steve: There's nothing you can do about that

Michael: You could put a queen excluder over the entrance. I think I've lost four frames of cells this year – I've never had that happen before. Maybe you lose one in a summer, maybe, not even every summer, and this year I've lost four.

Steve: Something's not right

Michael: Something's up. It was a swarmy year so maybe there's more virgins flying around the apiary. That cell builder is just crying out for a queen, and she smells that flying around the apiary, and in she goes. And they accept her immediately. It's not like a production hive, they accept her immediately, and she does her business – she kills all the queen cells.

On day five after grafting I reunite the cell builder with the queen right unit, above a queen excluder, and after five more days they're ripe, so it's like an incubator. It had to have happened in those five days between when I graft and when I reunite the colony. When you go to set up the cell builder you have to check thoroughly for any queen cells on the frames before putting in the grafts, because she will emerge first and kill them.

Sometimes when adding bees from brood factories [nucs] you grab the queen and put her in by mistake, so when you come to check the hive before putting in the grafts you see eggs and larvae in the cell builder. That's understandable. Sometimes you miss a queen cell hiding in the corner of a frame, so she emerges and kills the grafts sometime after uniting the colony. So I am absolutely fastidious about finding queen cells – I shake the bees off every comb and look for queen cells. They would be emergency cells. Do I ever miss any? Of course I do. But four in a year? Four cell builders in one summer? I'm sorry, I can't believe that, so I think they were coming in from outside, flying in during that five day period between grafting and reuniting.

Steve: So a queen excluder…

Michael: A queen excluder on the entrance I think would solve that. I made some entrance excluders. They look like an old queen and drone trap if you've ever seen one of those, they use to have these things a hundred years ago. It was a little unit you would attach to the front of the hive with an excluder on the front. Some of them have two compartments, with drone cones in one, so when the swarm tries to leave the hive the queen gets stuck. Well, have you ever seen one?

Steve: No

Michael: Ever used one? Are they for sale in the catalogues anymore? No. Why? Because they didn't work well enough. That's why things disappear. But I think in this case it might be beneficial. You've seen my cell builders and you know how strong they are.

I've started to second guess myself, and I wondered if the drones would clog up the excluder so the hive suffocates. Since the issue of rogue virgins entering my cell builders was so...almost non-existent, so I just never bothered. But now, with four in one season, oh my god.

Now I only do four cell builders, and each round it takes about three and a half lots of queen cells from my cell builders to populate my mating nucs. We've got 160 mating nucs to catch [i.e. find and remove the mated queen] every four days. We have four cell bars per frame, so 48 queen cells – some don't take so you have somewhere between 40 to 48. If a virgin destroys one of those cell bar frames, I'm short.

Sometimes when we put a cell in a mating nuc they don't accept it, and make a queen of their own. When we go to catch the queen, we find a virgin running around in there. So, the last time we lost a cell bar frame we went and caught those virgins, and introduced them into the mating nucs on the next round where we were short. That worked OK, but this last time we didn't have any virgins, so we had a group of 32 mating nucs with only 2 cells to put in. That's 30 mating nucs that I'm not going to be able to catch next time; it's going to put me short. What can I do? It's one of the things you have to deal with

Steve: Yeah

Michael: I don't think we've ever had such low catches on queens as we did this year. Every once in a while, we'll get one of those catches on one of the rounds where instead of 130 queens there might be 110. Well, the first catch this year was 109, we've had two that were 90 or 85. Can you imagine, 85 out of 160?! The ones that didn't have a mated queen either had a virgin or were queenless. Queen pupae, dead in the cell, never emerged. Fully formed, it's not like they died when they were larvae – fully formed but dead. I don't know, it's been a crazy year. Is it weather? Maybe. We've had crazy crappy weather this year.

Steve: If that's the thing that's changed, then it might be.

Michael: It is. I've done it the same way every year.

Steve: So it's been wetter than normal?

Michael: Yeah, rain. I'm surprised they had enough good days to actually get mated. I mean, the brood patterns look good, so we'll have to see how they last. Maybe next year they'll start to turn into drone layers or something.

Steve: What do you do with drone layers? Do you throw the comb away when that happens?

Michael: No, you just unite them with somebody else.

Legends from the beekeeping world

Steve: Have you got any heroes from the past or present, maybe some legends of beekeeping that you look up to?

Michael: Of course, lots. I wish I could have met some of the authors of the old books I read. Manley, I would have loved to hang out with Manley. I really would have, I mean, this guy, I think he was one of the best beekeeping writers ever. I love the way he writes. He says what he thinks. If he disagrees with somebody he says he disagrees with that person, he even names that person!

Steve: He does [laughing]

Michael: But it doesn't matter anymore, they're all dead! I'm thinking that way. I should say what's on my mind, and I should say I disagree or agree with people if I think it's necessary to do so, because someday it won't matter. I hope that people will read my work and understand. I'm not being an ass, just saying what I think.

Steve: Yeah

Michael: Yeah, for sure, Manley. Of course, Brother Adam – I wish that I could have met him. Some of the local beekeepers like Charles Mraz, I knew him, he was very good. We had another good beekeeper here, Ed Hazen; he was a comb honey expert, so he made all the comb honey for...everyone, the guy was incredible. When I was a young beekeeper attending an association meeting there would be Charlie Mraz and Ed Hazen, and some other old timers. Ed and Charlie were often opposites in their management. Charlie believed in never handling frames – he never handled frames, he handled boxes. Ed believed in total management, and re-queening and manipulation. Who made the honey crops? Ed Hazen. Who sold comb honey to Charles Mraz? Ed Hazen.

Steve: Right

Michael: I wish he was still alive today. I have questions now. I was too star struck when I was 25–30 years old, to come up to them and ask questions. Now I have questions, but they're all gone. It's too bad. It always seems to happen that way for me. There was a beekeeper in Montreal [Dr V R Vickery]. He was professor emeritus of entomology at McGill University. I believe he grew up in Nova Scotia. He wrote a book, something like, "The Honey Bee", and in that book he talks about wintering nucleus colonies. This was in the '90s and he was doing it before then. He wintered nucleus colonies in Manitoba.

Steve: Cold

Michael: Way colder than here. He put them in cubes, and wrapped them all together in a block, not as individual colonies, like what we do here. He had great success, and he believed that this was the solution to the closed Canadian border. Canadians used to buy package bees from America. At the end of the season they would kill all their bees, and take all the honey off

Steve: They just assumed they wouldn't make it (over winter)

Michael: Yeah. Then next spring they would buy packages from the South and re-establish their colonies. Just think, you wouldn't have to treat for varroa, would you? Then the Canadians closed the border to US packages, and now they're stuck, or they're buying packages from New Zealand. That's ridiculous. But he believed that the solution to the closed border was wintering nucleus colonies.

And it exactly is. There are plenty of places in Canada where you could winter nucleus colonies the same as we do here – anywhere along the southern border of Canada, anywhere. If you can winter a two-storey hive, you can winter a two-storey double nuc, no problem. Why they haven't taken up that challenge and winter nucleus colonies so they have their own stock is beyond me. You pay $250 for packages coming from New Zealand, and I understand it's pretty horrible stuff – how stupid is that?

There's another mite out there you know, they have it in Indonesia, called tropilaelaps. From what I've read it's worse than varroa. They'll kill your colony in summer. Yet they bring in packages from New Zealand and Australia. They already have apis cerana in Australia, the host of tropilaelaps – it's just a matter of time – then we'll have tropilaelaps in North America. It doesn't make a lot of sense. I don't see why people aren't listening.

Steve: Maybe they will.

Michael: Well I really hope so. When I started raising queens I didn't realise what would happen. In just a few years I was seeing amazing results, like, "wow, these are good bees".

Steve: If you hadn't done that maybe you wouldn't even be a beekeeper now

Michael: Maybe I wouldn't, maybe I would have lost it already. I was struggling that whole time, trying to pollinate apples and make a honey crop. Man, that's difficult. Not only what pollination does to the bees, and the beekeepers and the equipment, but every time you pick up a hive and move it you lose honey crop. If I'm picking up an apiary and moving it 15 miles to the orchard and back two weeks later, I'm going to lose at least a medium super of honey with that [per hive].

So, what's that, about 40lbs of honey, and it's $2.50 bulk – there's $100. Pollination doesn't pay $100 for apples. It pays $65 to $75. So on every colony you move you're losing $25–$35; how stupid is that?

Steve: You mentioned Brother Adam?

Michael: It's like Vickery, I found out about his book, read his book, I tried to get in touch with his family – he died a month before I read his book. I didn't get to meet the guy. This is typical. Yes, I would have loved to meet Brother Adam, and his long-time assistant, Peter Donovan, but he died two months before I got there. Put it this way: when my ship comes in, I'll be at the airport! That's my life [laughs].

Steve: Is it more about comparing notes and chatting, or do you have particular burning questions that you want to ask?

Michael: I do have questions, sure, but nobody can answer them now. That's one reason why I enjoyed meeting David Kemp so much, because I had questions for him about Brother Adam's cell building set up.

Steve: And he was able to answer

Michael: Yes, but you know how it is when you have a million questions, but you only remember four or five?

Steve: Yep, that's why I have mine written down!

Michael: So I have a million more questions for him. One of them would be, "how many times did Adam re-use the cell builder?" Also, "does the acceptance of the cells go down as the cell builder gets stronger?" – things like that. I hope to be able to ask him some day. I wanted Brother Adam's yard sheet. Did Brother Adam use a yard sheet for selection criteria, did he do that? He must have. But it's all locked away in a vault and you can't get access to that stuff, because Buckfast Abbey has got a problem with it.

Steve: You visited Buckfast Abbey didn't you?

Michael: I did

Steve: And it was a bit disappointing?

Michael: Yes. To see his mating station abandoned and rotting? My God.

Steve: Why did you hold Brother Adam in such high regard?

Michael: Because I think he's got the best cell building plan. He showed that he could take a group of bees, and select for, say, resistance to Acarapis woodi – he has shown that you can do that. And I think with his cell building plan you get the best queen cells.

Is there any perfect beekeeper? No, of course not. Everyone has issues. Right now, I talked to someone who's running the bees at Buckfast Abbey and they talked about how horrible it was that Brother Adam used sulphur thiazole to control American Foulbrood disease. So what? So, this is going to totally overshadow everything he did in his life? I think that's wrong. Everybody makes mistakes. I don't think it should outweigh everything he ever did.

Thoughts on buying package bees

Steve: It's interesting about the old beekeepers, and how some things don't change. You're quite interested in the old books aren't you?

Michael: Yes I am. What we are talking about today, wintering nucleus colonies, it was going on before package bees. You know that ABC book I showed you this morning, the one published in 1891? I can find a reference in there by CC Miller, where he wintered nucleus colonies – double nucleus colonies where if you put the boxes together they form one cluster where they touch, with the wall being a divider between them. That's amazing, because that's what we discovered just by doing it and seeing them, but he wrote about it in 1891.

What happened? People forgot. Why? Because the package bee industry came along and was good for a while. You could buy a package of bees, in catalogues, through the mail, cheap. My first ones were $10.50 delivered. It was a good product, and if you wanted them May 1st you got them May 1st. You can spend $150 on a package of bees now. You order them for May 1st and you might get them May 21st. It's different.

Steve: This is what I don't understand, because if they are selling a crappy product it can't last, surely?

Michael: Wouldn't you think? But if that's the only access, and you want them badly enough…plus I think with a lot of new beekeepers when they buy a package of bees and it fails they think it's their fault. They may have done something wrong, lots of beginners kill their bees every year, but there is an inherent problem with package bees. It was there when I started, and the problem is there is an imbalance in the bee population in the colony.

Let's say the package provider shakes the bees into the box on Monday, and they ship them on Tuesday, or the re-seller picks up a big order of a thousand packages on Tuesday. On Saturday morning the beekeepers pick them up and put them in their hives on Saturday afternoon or Sunday morning, and the queen comes out of the cage on Tuesday or Wednesday. Now, it's been more than a week. The youngest bees in that colony are more than a week old. Then she starts to lay, so there's no brood emerging for 21 days, and they get more and more brood, more and more larvae to feed – more and more nurse bees are needed, at a time when there are fewer and fewer nurse bees in the colony because they're getting old. Pretty soon older bees have to behave like nurse bees, which is an issue, and they supersede at around 3 weeks.

It's always been there and it's the inherent problem with package bees. Even when package bees were great that still happened. The way to deal with it is to open the hive after ten days and put in a frame of emerging brood from some other colony, so they get their nurse bees, and that overcomes that 3 week supersedure thing. It's very easy. People think it's the queen, but it is not always the queen. Sometimes it's just that inherent problem that packages have.

Steve: Interesting

Michael: So 30% are going to fail right off the bat, because once you have a supersedure in a package, unless you have a long season you're never going to get the strength up. Plus what happens when the virgin flies out and doesn't come back?

When I started some of the package bee producers were asking for queens back. If you ever got an exceptional one they wanted that queen back, to use for breeding. Nobody does that anymore. It used to be that if you had a failure you could call them up and they would send a new queen. They hardly do that anymore.

What do you do if you have a drone layer? Even if they gave you a new queen it's too late for that colony, because they don't have any worker brood. The new queen is going to get slaughtered, so that colony's doomed.

What happens up here is that people have to buy packages from a re-seller because the producers won't ship to so far away. There's people making a whole lot of money for simply buying a batch of packages and driving down to get them, then driving them up North to sell. This guy I know bought 24 packages of which 12 were drone layers. That happens, it can happen to anybody in bad mating weather. So, he calls up the guy he bought them from in Massachusetts, and the guy says, "I don't know what to tell you, I didn't grow those bees, it's not my fault." So, he calls up the company that made them, "I didn't sell them to you." What's he supposed to do? It's outrageous. You'd think with service like that they'd fail, but they sell hundreds of thousands of packages each year.

Steve: As your message spreads are you making enemies?

Michael: Probably.

Steve: But there hasn't been…

Michael: I haven't been attacked!

Rob, Kate and Zach

Nucleus colonies

Queen introduction
and new truck

It's not just packages. There's a major queen rearing company, I understand they make 1,000 queens a day, and the Apicultural Research Professional and Lab Manager bought about 250 queens from this company. They installed them in colonies and only something like 10% of them got accepted. Now, she knows what she's doing. Everybody makes mistakes and suffers loss, but 90% failure?! So, she called up the company to ask if they would replace her lost queens, and they said no, and that she must have done something wrong.

She called me up to ask if I had any queens, and she called another guy in Oregon, so I sent her what I had and she got 100% acceptance of my queens and 100% acceptance of the other guy's. She wrote an article about this in a bee magazine without naming the company concerned.

Well that's outrageous, you know? It's not like some dink bought a bunch of queens and blew it, you're talking about a university and these people know what they're doing, and they got results like that, terrible. They could have afforded to replace a couple hundred queens.

Slipping standards

Steve: It's funny because when we come to America from the UK we're always quite impressed with the level of customer service, and people always seem very welcoming and helpful. What you're talking about seems un-American in a way.

Michael: Yeah, well customer service has gone down the tubes with a lot of companies.

I've got a real problem with my extracting equipment – my wax melting, separating cappings from honey. We use to use a shovel. We'd shovel the cappings into the wax melter. It takes a long time and you can burn the honey, so I bought a capping spinner and an auger to go with it, both from the same company. The capping spinner is supposed to do 10–15 drums of honey a day. As I told you yesterday I only need to do six. The auger has a hopper with it, so the cappings go into the hopper, through the auger and into the spinner. I was all excited, I spent thirteen grand on the equipment, three thousand for the auger, and it worked great. Honey's coming out into the sump, wax is falling into a pan at the side, "Wow, look at that, it's dry! It's powder!" Unbelievable.

Then it stopped working. It slowed down, then overflowed, so we had to get a gate valve to control the flow of honey out of the extractor, then it clogged up and stopped. I called the company up. He said I was pumping it up too high. I started talking to people and they were saying they pumped it up 6 feet; I was only going 30 inches. So, I took the auger apart and saw that the four holes that slurry goes through were becoming blocked up. I rang the company to tell them what was happening. They came over and cut the auger pipe shorter, so that I could easily get to the holes and unblock them if needed, but it would block up straight away. Then I lowered it to 20 inches. I tested it with some clean honey and as soon as the honey hit the auger blades it would groan and slow right down. I called the company again, had to leave a message, but they didn't call me back.

Finally they asked if I tightened the clutch. I said, "tighten the clutch?", and he said, "yeah, we don't tighten the clutch all the way, we're afraid people might cut their fingers off." Really. I'm not going to stick my fingers in the damn thing. Oh, and before that, before they fixed that, I noticed a bushing sitting in a puddle of honey in the pipe. There's only one way that could have happened: they put the bushing holder in backwards, so it was on the wrong side and it fell out.

There's something wrong with it. Only a quart of honey and it slows right down. It gets so hot between the motor and the shaft, it's not supposed to be that hot. I could stop the whole thing, motor and everything. It wasn't slipping. If something gets jammed in there the motor will go but the clutch will slip, that's what it's designed for. No, I could stop the motor. They didn't believe me so I sent them a video of it from my phone. They didn't answer me.

A friend of mine who used to sell their equipment in a shop said that they had showed him my video the other day, and asked him what he thought was wrong with it. They wouldn't even talk to me.

Steve: No

Michael: So, I don't know. We put 32 tons of honey through that damned thing. I had to sit there on the stairs next to that auger the whole time, unplugging when it blocked. I'd ring them and they wouldn't answer me. Finally their solution, "send it back to us and we'll give you your money back." Well, that's not going to work, I need the auger to use the $10,000 spinner I bought from them. Nothing. So at the end of the season, in January they took it back, and said they'd be in touch. I waited until April, they still hadn't contacted me, so I called, couldn't get anyone to speak to me, I left messages, nothing.

So, I rang accounts receivable. Guess what, somebody answered. Of course! Of course they'd have somebody answering that. I was a little rude, and asked to speak to the owner or the manager but was told he was too busy to talk to me. He just said he'd send me the cheque on Monday and hung up.

Talk about customer service. This is typical now, typical. I ordered $7,000 of foundation and they told me I'd have it in a week. A month later, "when are you going to send me the foundation?". "Oh, we didn't have it all in stock, we were going to ship it all at once." Well, why tell me it would be a week, and why not call me to tell me there was a delay? "What should we do?" "Ship me what you've got!"

It's just…it's unbelievable, and it's every industry

Steve: Not good

Michael: Not good.

Steve: Sounds a bit like my country [laughs]. This is the thing isn't it? All you have to do is be competent, and decent, and you stand out because everyone else is so bad. We had that in our business.

Michael: A guy I know found an auger for sale on Facebook, so I may be able to get one from somewhere else that works.

Steve: OK. Time for a cup of tea.

Michael: There you go

Steve: Thank you.

Some notes and diagrams on Mike's sustainable beekeeping method

I really enjoyed that interview and my time spent in Vermont. It is obvious that Mike Palmer has a message for the beekeeping world, a powerful message based on personal experience and many years of learning what works and what doesn't. The message is, "Don't import queens, or packages or nucleus colonies from far away – grow your own queens, raise your own nucs, and you will see huge improvements very quickly." He knows that not every small beekeeper will be able to run a queen selection and breeding program. Those people, he says, should find a local beekeeper who is doing that, and get queens from them.

In Diagram E I have tried to illustrate what Michael Palmer calls his sustainable beekeeping method. The key idea is to use nucleus colonies to provide brood frames for creating cell builders, boosting production colonies and making up other nucs to be sold or taken through the winter. The mating nucs are quad boxes containing sixteen half-length shallow frames (four in each section). A movable frame feeder acts as one divider and perpendicular wood dividers slot in to create four mini hives from the one shallow box. Mike takes his last queens through the winter in three of the quad boxes stacked together with the dividers moved aside. What is home in the summer for twelve queens becomes home for one in winter. Most survive the harsh Vermont snows and are a valuable resource in spring.

Mike's queen rearing method was inspired by Brother Adam. He starts a cell builder by putting a box of honey over a queen excluder then a box of sealed brood above that (see diagram A). The sealed brood is harvested from the nearby nucleus colonies. He inspects for queen cells every few days and removes any found. After ten days the hive is moved backwards and rotated 180 degrees (see diagram B). A new floor is put down where the hive was, and onto this goes the honey super that was above the queen excluder followed by the box of emerged bees, then a syrup feeder and a roof. Mike removes two central frames from the box of nurse bees and replaces one with a frame full of pollen. A space is left for the frame of grafted larvae, which is inserted later in the day (see diagram C).

Diagram A
Starting off a cell builder - add box of sealed brood

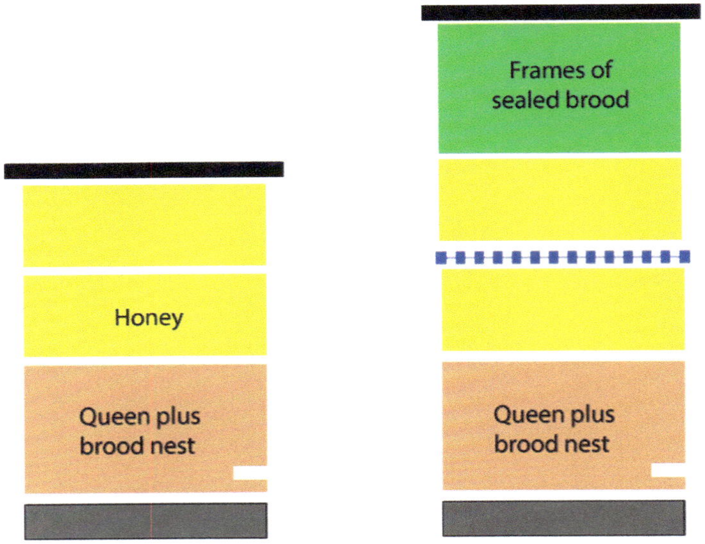

Diagram B
Ten days later - split into two hives

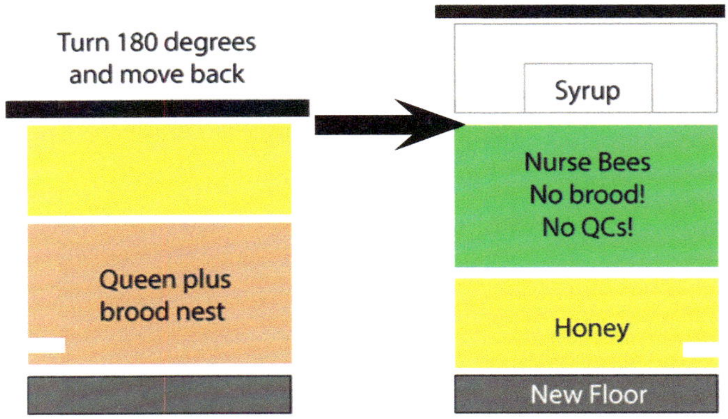

Five days later the cells are sealed and Mike re-assembles the hive as shown in diagram D with the cells in the top box above honey supers and a queen excluder. The cells will be taken to be placed in mating nucs shortly before the queens are due to emerge.

Diagram C
Grafting day - add pollen frame and grafted larvae

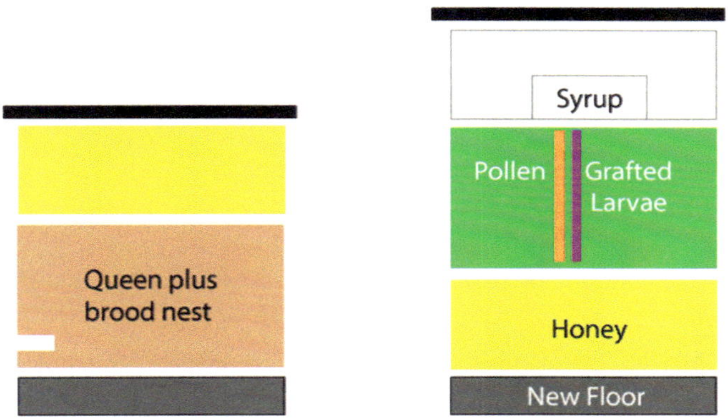

Diagram D
Five days layer - put hive back together

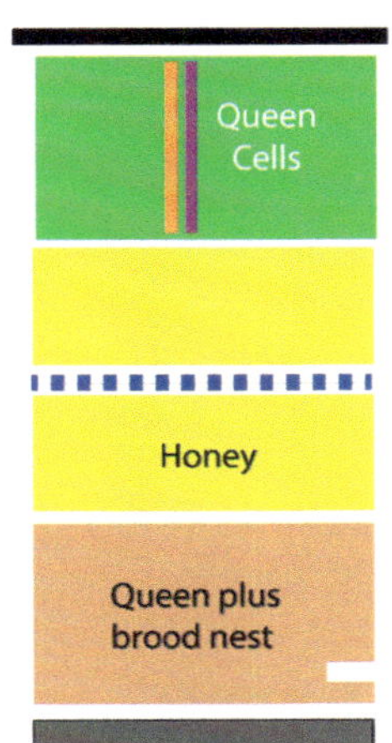

The whole process is then repeated over and over throughout the summer. This method produces queen cells which have an abundance of royal jelly and is, as far as Michael Palmer is concerned, is the way to get the best possible queens.

Diagram E
Sustainable beekeeping

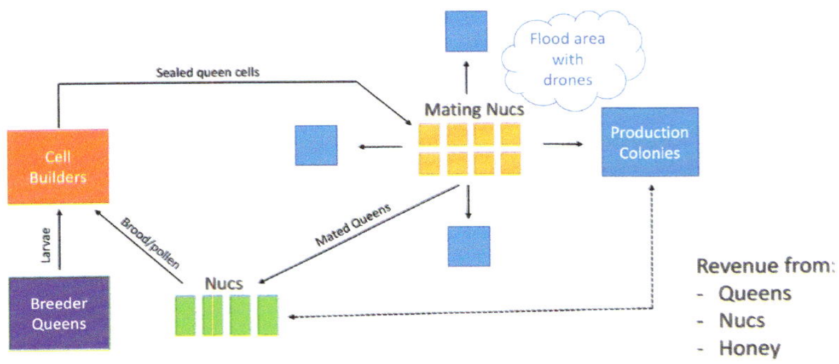

Mike gave me a mason jar of his honey and a tub of comb honey to take home with me. As he dropped us off at the airport and said farewell he gave me his "Vermont Beekeepers Association" baseball cap, a treasured possession, which sits on my desk next to me even now, as I write this.

Ray Olivarez, Orland, California, USA

'The thing that I've learned is you can train but until you experience it with bees, I don't know why it is, but you just have to live it. There are so many variables and you never quit learning. If anybody tells you that they have it all figured out they are so full of baloney.'

Introduction to Ray

The beekeepers in this book are very busy people. When not running their beekeeping operations they are usually in high demand as speakers at various events and conferences around their country, or indeed the world. It took a fair bit of arranging to get a date which worked for Ray Olivarez but eventually we settled on Tuesday 23rd October 2018 at the Olivarez Honeybees headquarters in Orland. The following day was to be my day with Randy Oliver. I was not expecting to be at my brightest, but this was my chance to get interviews with two of the giants of the beekeeping world, and I was not going to miss it. Olivarez Honeybees is a large company that is still very much a family enterprise, which appeals to me.

My flight from the UK landed at Sacramento International Airport (SMF) on Sunday night, so I had the Monday to try to get over my jet-lag, before making my way to the car hire company on Tuesday morning. Yes, I was going to drive in America! Driving in a foreign land, on the other side of the road, with road signs and customs I was unfamiliar with was something I had prepared myself for, but I was out of my comfort zone. Travelling on my own to meet people that I don't know, and who don't know me, to interview them for my book is, in fact, several zones along from comfortable.

The process of collecting my rental vehicle turned out to be entertaining and profitable. The young lady who took me to my car was distressed to find that it did not possess the satellite navigation system that I had ordered. She apologised, went away, and returned with the keys for a much smarter looking motor. It was something big and comfortable and American, but to her horror, this too was missing the vital piece of navigation hardware. She found more ways to humbly apologise, which at one point involved complimenting me on my sports jacket (a rather fetching Harris tweed), and this time she handed me the keys to a black Jaguar XF 3.0 V6. I wasn't going to hand them back, especially as I was charged nothing for the upgrade, so off I drove, rather pleased with myself.

Within minutes I had taken a wrong turn and had to pull over in an airport car park to compose myself. How the hell am I going to get to Orland in time, thought I, if I can't even find my way out of the airport?! Family members who have experienced my heightened stress levels when driving on holidays abroad would have recognised the tell-tale signs of a man about to explode; the beads of sweat forming on the brow, the shortness of breath, the waving of arms and fidgeting of arse.

Once I got going again, having realised that the navigation system was the same as the one on my Land Rover back home, I found the roads of California to be fantastic. My whole "driving in America" experience was relatively straight-forward and enjoyable. The potholed crowded streets of the UK are a nightmare compared to what I saw over the pond. I may have broken a speed limit or two in my fancy V6, as an Englishman abroad is undoubtedly entitled to do, and I arrived at Ray's place in Orland an hour early.

Ray introduced me to his wife Tammy and various other family members and business colleagues, who were all about to have a board meeting. I gave them a gift of a bottle of mead that I had brought with me as I know they are interested in this historic liquor made from honey. I left them to it and spent some time with the impressive Ali Churiel who looks after what Ray calls "the bible" (the hive report). Her job is to ensure that the central record of all the hives in the whole company is up to date and accurate at all times. It is a prodigious task requiring diligence, attention to detail and dedication.

They were in the process of bringing 10,000 colonies back to California from Montana, and the locations and condition of every hive were recorded and tracked using the hive record. Ray later told me that he sends many of his colonies to sites in Montana to escape from the intense agriculture and more stressful environment in California. This year he kept 5,000 back in California, and the rest went to Montana. They get sent on their summer holidays to the gorgeous, pristine countryside where they can make honey and have access to wholesome natural pollen, and recover from the rigours of almond pollination. Ray's sites in Montana are protected by law; no other beekeeper is allowed to set up within a 3-mile radius of his bees, which is not something that you find in California. The bees return in October broodless, because it gets cold up there, which provides an ideal opportunity to treat for varroa mites.

My interview with Ray took place in the boardroom after their meeting had concluded. He had only just returned from speaking at a Bee Culture event in Ohio the previous day and was off to Idaho later that evening, so I think he was probably feeling as addled as me. Afterwards, he gave me a tour of his facility and showed me the mountains of queen mating mini nucs that were being cleaned up and stored for next year. He had a giant chest freezer packed full of bags of pollen which would be used at some point to feed his bees. Ray told me that proper nutrition is an important part of keeping his colonies healthy and strong.

Cleaning up mating boxes

Pollen in the freezer

Talking with Ray

Operations and the business of bee-farming

Steve: I think you were just talking to your team about new software or something?

Ray: We are having to change our management practices because it's getting more and more difficult to keep the bees alive. There are timings that you have to keep to; varroa mites don't wait for you, the honey flow doesn't wait for you, queen cells do not wait for you. There are timelines that you just have to stick to. Database software can also help you in the future by looking at your metrics. What's successful and what isn't? If we do this, in this amount of time, it looks like we get a consistent 5% increase every year...oh, OK, we need to roll that out.

Steve: Some of it is driven by your ambition to keep growing I guess? Your numbers are already huge. Are you just planning to keep growing and growing?

Ray: No, not necessarily. We're growing more vertically. We produce honey. Instead of selling honey to a middleman, at $1.83, we're starting to pack our own honey. We've got Lyle as our Director of Marketing, and we are getting $6 per pound after all of our costs of packing – that's huge. So our goal is to diversify more. We are already diversified; we raise queens in Hawaii and California, we pollinate, we produce honey and we do package bees.

The revenue streams we can get from what we do helps us, because there is less pressure to hit certain numbers on the bees every single year, and push the bees. We're trying to take pressure off the bees, because it's pressure and stress that causes problems. The more stressed the bees get the more susceptible they become to viruses and other problems.

If we're going to pack our own honey we need to be more consistent. We don't want to be selling to K-Mart and Walmart and Costco because the profit margins are not very good. In order to produce a quality product we have to go back to accountability, standard operating procedures, and really dial in the processes, and then the culture of the company; the mindset of the employees. The better all that gets, the better the quality of the product gets.

We raise a lot of queens, and it's like breathing to us, but some people just can't get it – they can't even produce a few thousand queens. We cross train our employees to provide depth. They don't just understand their little bit of the process but the whole thing, so they know why they are doing what they are doing and why it's important. It's a big interconnected organism. Our business is not a text book company and nor is our industry straightforward.

Steve: It's farming isn't it, when you're out in the yards? Then when you get back here it's something else; business processes and systems and whatever.

Ray: We are even more affected by the weather than most farmers. If you get a chance look up Ian Steppler. He might be another good one for your book.

Steve: What can you tell me about him?

Ray: He's a Canadian beekeeper, but he has a tremendous following, because he does a lot of blog posts and has a you tube channel.

[Got distracted by staff, eventually moved to private room with just Ray]

How it all began

Ray: I'm a second generation beekeeper. My Dad started in the bees in 1965, and he started working for a guy called Clarence Wenner. Northern California has a lot of commercial queen producers because of our area, it has a Mediterranean climate. The first thing we're doing is pollinating almonds, sometimes as early as February 1st, which is still winter. We're in the low 60s to mid 70s by then.

The whole history of Northern California beekeeping is about raising queens. You had UC Davis which had the Bee Research Centre, which was one of the best in the world at that time. A guy named Laidlaw was there and he taught a lot of the beekeepers about queen rearing in this area. You really had three big beekeepers back then, and since then everybody has kinda branched out from them. You learned from either Wenners, or CF Koehnen & Son or Don Strachan.

There were a few points in time that were critical; in 1987 varroa mites

Keeping track of 16000 hives

Halloween at OHB

Trucks for moving hives

Larger incubators

showed up, before then it was tracheal mites. Probably the biggest event that changed beekeeping in the United States was the almond industry...

Steve: When did the big change happen?

Ray: Oh, there were some numbers on that the other day...I'm trying to remember...I'm going to say around 2005–2006. I think around then we were at about half a million acres of almonds, but you can check those numbers – Randy probably has that. Now we're at 1.3 million acres. That requires 2.6 million colonies for pollination and there's about 3.2 million colonies, maybe, in the whole of the United States. Almond pollination is a singular event, it's just so large.

Ray: My Dad started working for Clarence Wenner. There was a lot of interaction between Canadian beekeepers and California beekeepers, because of the queens and package bees. Back in the 50s and 60s and 70s and early 80s most of the queens and package bees that were produced in Northern California went to Canadian beekeepers, because Canadian beekeepers would gas their bees at the end of each season.

Steve: Yeah, I heard about that

Ray: They would re-stock them every April. In 1987 that came to a screeching halt which almost put my Dad out of business. It put many Canadian beekeepers out of business too.

Steve: Was that because the government wanted to prevent imports?

Ray: Yes, the Canadian government. We're on the same land mass [laughs] – you know, there's this imaginary line that's the border and the bees do fly across. The thing about it is that when the border was closed just as many or more queens were smuggled across the border, because they needed them. Eventually they started getting queens from Australia.

They started getting packages out of New Zealand and it's actually from New Zealand where we think we got the varroa mite from on the Big Island of Hawaii. As they shipped the packages from New Zealand to Canada they landed on Oahu, unloaded the packages then loaded them back onto another plane for their onward journey. We fought against that but didn't get anywhere. We are pretty sure that's how the varroa mite got there. We don't have tracheal mites on the Big Island, but I don't think we'd have it anyway – our stock is resistant to it, most stocks are – you rarely hear about tracheal mites anymore.

But my Dad got started in 1965 working for Clarence Wenner and then I think in about 1968 he bought out a gentleman named Art Banta of Los Molinos. Art was a beekeeper originally from Davis, California then he moved up to Los Molinos, and was a very good queen breeder, incredible; in fact his son-in-law was Cliff Thomas who is known all over the United States. He's passed away now. My Dad grew his business in Hamilton City then eventually we moved to Orland. What he was trying to do was give the other guys some room.

Back in the day when there was 'bee etiquette' nobody would set up a facility within at least 3 to 5 miles of another beekeeper, and you tried to keep your bee yards a couple of miles from each other. Now in California it's a free for all. I've got a beekeeper who set up a quarter of a mile over here [points] and I've got another one right over the overpass here, and they don't care. We've had to change what we do here because of the number of bees that they keep at their places, so we're buying properties to raise our queen cells in a different spot, to give us some room.

Steve: Is there plenty of room?

Ray: No. The way we look at it, instead of fighting it and spending all of our energy being upset about it, we focus on what we've got to do in our environment right now. Beekeeping is very territorial, even more so in my Dad's day. Beekeepers worked a little bit with each other but they didn't share information very much, they competed for customers of queens and packages and pollination. We don't have to do that anymore. You need to work with your neighbour; your bees are as good as your neighbours bees. The more information you can share the better the environment will be, hopefully.

We keep an open mind. I don't hold anything against him. It's made our business a little bit more difficult but it feels to me like for the last ten years the more difficult beekeeping has got the better we have done.

I'll go back to the history; I go off on little tangents every now and then – sidebars

Steve: Sure

Ray: My Dad started with 100 hives then he bought Art Banta out and I think he got up to about 2,200 hives by 1982. I was a senior at that time. In that spring of 1983 I was a swine herdsman at Modesto Junior College – I was going to be an agricultural teacher. I was allergic to bees.

Steve: Are you still?

Ray: No. I grew out of it when I got into the bees. I always worked building boxes for packages, wiring frames and I was around the bees, but I did not want to get stung because it was anaphylactic right away. I took shots for about 3 or 4 years.

In 1983 Dad moved his bees into the almonds and it was the first time that we became aware of what an El Nino is. It's a lot of rain! Most of my Dad's bees were pollinating orchards along the Sacramento River. In three nights he lost almost all of his bees down the Sacramento River to the flood. We moved them to 100 year levies which were still near to the orchard, because he was getting paid, but when we came back the water had gone 5 feet over that. It was a devastating flood event.

From that point he never really recovered.

Steve: Well, no. Gosh.

Ray: When I came home to help my Dad I never went back to Modesto. I went to Butte College to continue my education, we bought a beekeeper out with 500 hives, but it wasn't quite enough revenue for two families. I was married to Tammy and we had some young kids, we had varroa mites show up and it was very difficult. I went to work as a surveyor.

I kept 100 hives and grew them to 1,200 hives while I was working full time surveying until I had to decide to do one thing or another, I couldn't do both. I chose the bees. Then, around 2006 I took on a business partner named Dan Cummings. He was working for the Violich family, Capay Farms, who are a large almond and walnut producer. He is now their CEO. Originally they were wanting to partner with a beekeeper to secure pollination supply, because he could see what was happening – the almond industry was going to take off. There was a shortage of bees and the bees were not healthy for the most part so they wanted to have their fingers on the pulse. They saw my work ethic and thought, "let's grow with Ray."

Dan wanted to become a partner so we went 50/50 and we had a partnership for about ten years. In that time we went from 1,200 hives to about 10,000 hives.

How we grew the business

Steve: So how do you do that? Is it just about keeping on splitting?

Ray: Yeah, we could split, but in order to grow you need locations. To get the locations we bought beekeepers that were retiring.

Steve: Ah.

Ray: That's how we did it.

In the meantime queens is what my Dad did, and package bees; raising queens for us was like I said earlier – like breathing. It's just second nature. We started raising queens, and of course the package bees are a management tool; we shake the bees at their peak so that they don't swarm on us. During that time we also obtained some locations in Montana for honey production.

Steve: Whats the forage in Montana?

Ray: Well where we're at its mostly dry land, farming, so a lot of cattle ranches, and we're surrounded by the Custer National Forest. We don't have the irrigated alfalfa but what we do have is just a pristine environment. The idea to go to Montana was to get the bees out of California.

We grew from 2,000 hives. When there are five or six of you working on 2,000 hives you learn more about beekeeping than you do if you are working for a big beekeeper with 12,000 hives. They are not thinking, the general employee, about why they are doing what they are doing. We had a long and steep learning curve.

Beekeeping is not something where you can read a book and then do it. You have to live it and experience it. I hear all the time beekeepers saying, "God, I can't find anybody like me – that knows and sees what I see." That's because they took it from one hive to 1,000 hives or 10,000 hives over a long period and they made all those mistakes and learned from them, you know? It's even the case with beekeepers' children coming into the business.

We didn't have plans for our kids to work in our business. They were going in different directions but because of the scale of the business here and the need for bees there is the potential to do very well. It's supply and demand. We see a lot of our friends who have kids that were lawyers or teachers or firemen, it doesn't matter what, coming back to work with bees. Twenty years ago there was nothing to come back to. "I'm not going to let my kid become a beekeeper, I'm not going to let them work this hard!" Now they're coming back to the bee industry, it's really interesting. I never dreamed that I'd see that.

Steve: All of that's driven by the almonds isn't it?

Ray: Yeah, there's a lot of things.

Mentors

Steve: What about mentors or heroes?

Ray: I had a lot of mentors. Of course I learned a lot from my Dad. When I was younger I was a typical son; Dad thinks he knows everything...

Steve: And you know better?!

Ray: I know better, yeah. He let me explore those things and make mistakes. Every time I asked him a question he had an answer. I've learned with my kids to listen. Of course, we're learning as a family to do that.

Leonard Pankratz has been a great influence on me. He is a beekeeper that originated in Canada but when the border closed he moved to California. He used to have bees in California and Canada.

Pat Heitkam is another person that I get a lot of wisdom from, you know, just life. Pat's always been more democratic about things, getting people to work together and get along with each other so I've been inspired by him.

David Powell, who I've got to know better over the last five years, is about 72 or 73 now, and he got started around the same time that my Dad did. He is an incredible beekeeper with so much knowledge. There is also Valeri Severson of Strachan Apiaries. She watches over Tammy and I.

Kevin Ward is originally from New Zealand. He married Laurel Hill, who's father Oliver Hill was an incredible beekeeper out of Willows, CA. I talk to Kevin two or three days per week. When we are really busy we just bounce ideas off each other and I think I centre myself after I talk to Kevin [laughs].

Steve: Well, you need that, I guess.

Ray: Then there is Randy Oliver, who has a wealth of knowledge. He continues to become a better beekeeper. He knows the science, but he asks me a lot too, about commercial matters. His boys have finally come into the bees. Randy does mostly speaking and research. I've got a ton of respect for Randy.

What else did you want to ask me?

Advice on becoming commercial and challenges involved

Steve: OK, it's a whole different ball game in the UK to over here. In the UK if someone has 100 hives that's getting big. The biggest in the UK has 3,500. But it doesn't matter where you are, if somebody is keeping bees as a hobby and they think, "Hey, this is great fun – I'll quit my job and be a beekeeper!" – what advice do you have for somebody like that? Advice for the person who has grown and they have to decide whether to become a commercial beekeeper or keep it as a hobby?

Ray: It's changed a lot. If you want to take the current situation with the varroa mites and all of the viruses, the economy, what's available as a workforce, I think what you want to do is start off with maybe 2 or 3 hives for a year and just see what you can keep alive. What you can over winter will tell you how well you did with your bees.

Mother nature can make you look like an incredible beekeeper when she helps you out. You have to recognise that certain years are better than others.

This year we made 1,400 barrels of honey in Montana; we got good moisture and the flowers were there, so we were a lot better beekeepers this year! The bees did have to be good in order to take that opportunity – you could have all the forage in the world out there and still get low honey yields. I've known people like that, who kept going longer than they should have done because the area where they kept bees was so forgiving. The bees did it on their own for the most part, until it got so bad that even if they worked a little harder or tried to become a little better, it wasn't enough for them to stay in business.

In every region it's different. A hundred miles from here it's different than here. For us, right now, our biggest challenges are the mites, nutrition and of course, labour; having good people and keeping them. How do you retain those people?

I'd say if you could keep 50% of your bees alive consistently for a couple of years then maybe jump into more. I would expect of myself, if I have 100 hives and that was all I was doing, that I could probably put 90 into the almonds each year. We work on those ratios here, so if we can put 90% of our bees into the almonds then we are very successful.

Steve: That's all about getting them strong enough in time?

Ray: It's keeping them strong enough, keeping them healthy to go through the winter and into pollination. From that point you have time to make a mistake and fix it, but not before. Lets just say almond pollination is February, once we get to October if the bees aren't healthy then you are not going to make it through to February to make the money. If you can't make it to the almonds you won't have the money that you need to pay for everything. Nobody gives you anything for free! If you make a mistake in May you can still fix the bees by October, but if you don't have healthy bees in October they just dwindle and die out going into the almonds. That's why you hear about these 30% – 50% losses; it's occurring in October, November, December and January but it was caused by something that happened back in July, or didn't happen.

Steve: Probably varroa?

Ray: Yeah, then viruses. These viruses are in the bees but varroa allows them to express themselves. When we went to Hawaii we didn't have varroa mites yet. We sent samples to the bee lab in Maryland, and David Wick had a machine that separated particles, and we were able to detect virus loads. We had all of the viruses in Hawaii but we just didn't have varroa mites, right, we didn't have the stressor that causes the viruses to become a problem. We had Nosema Cerannae in Hawaii already, high levels, but when we got the mites all the problems of the mainland came down on us, plus small hive beetles. That's

OK because we have learned how to manage them and our breeding program over there has all of this stress; we've got varroa mites and viruses and small hive beetles, which helps us with our selection of breeder queens.

On queen rearing and instrumental insemination

We raise queens all year round over there which is one of the reasons we went to Hawaii. We bought Big Island Queens about 12 years ago because I wanted queens from Hawaii but I could never get them because breeders were sold out. Having that operation in Hawaii for ourselves has been great. We use probably 3,000–4,000 queens on our own hives here in California that we produced over there.

Steve: How many queens would you say you make a year?

Ray: This year we produced about 370,000 queens.

Steve: Thats a lot!

Ray: Yeah. We produced about 270,000 in California and not quite 100,000 in Hawaii. It's usually 80,000 to 100,000 queens in Hawaii.

Steve: Do they let you move queens from one area to the other?

Ray: On the Big Island you cannot take any used equipment there and you cannot ship bees there. You can ship bees and used equipment off the islands but not the other way.

Steve: So what, do you transport the semen or something?

Ray: We maintain our breeding stock using instrumental insemination.

Steve: You bring in semen from...

Ray: From here in California, from Sue Coby (New World Carniolans).

Steve: They don't have a problem with semen coming in as long as you sign all the forms and so forth?

Ray: Right. What we're worried about is getting Africanised bees over there so the government is working on some more concrete regulation, because they could arrive in containers or as semen. It's kind of loosey-goosey right now, it's kind of scary. They don't want to lose us as queen producers on the island, because it's a big source of revenue. They are working on guidelines for the importation of semen. Now for Canada, we ship a lot of queens there, and every 30 or 45 days we have to do a DNA test on our bees.

Most people that excel in bees love all bees and they love the work. I don't know much about the UK to be honest with you, but I think you mostly have the same issues.

Supporting hobby beekeepers

Steve: Well, the commercial beekeepers that I have spoken to in the UK all work crazy hours but they love their bees. Some people think that commercial beekeepers are just moving boxes around but all of the people I have spoken to care about their bees probably more than the hobby beekeepers.

Ray: Yes. One of our problems here is colony collapse disorder. CCD is an umbrella term for many things; it's the lack of forage, it's the varroa mites, it's the viruses, it's the poor beekeepers, it's the environment, it's everything. We have a hobby day here when almost 2,000 people show up. Only about 1,000 of those want to have bees, because they want to help the environment, and we know that 60% of them are going to fail because it's not easy. Those bees get infested with mites and they spread to our commercial colonies and make it more difficult for us, but they are going to do it regardless. If we tell them that we can't sell them packages they are going to find them somewhere.

We have that day when we bring in Randy Oliver and the tech transfer team from the Bee Informed Partnership, Larry Connor to talk about brood diseases, Kim Flottum just to talk about how to be a hobby beekeeper, we do demonstrations and we try to give them as much information as we can to help them be successful. It usually takes 2–3 years of them coming back before all of a sudden they start getting it. Then they don't buy any more packages from us because they are keeping their own bees alive.

Mite bombs

Randy will probably talk to you about this when he sees you, this thing called mite bombs.

Steve: Yes

Ray: We are seeing this even in our own colonies; there's always 2%–3% of the colonies in every yard where, for some reason you can't get all the mites, or the pheromones in that queen are super strong and maybe they are attracting mites. It seems that colonies that are stressed abscond and maybe look elsewhere for help, and they spread the mites. Randy has this experiment where they tagged bees and he's finding them even a mile away in other hives – that's how much these bees drift. That's a real concern, but again, we just need to keep working on that.

Steve: Regular mite monitoring?

Ray: Yes, regularly use alcohol wash. When our own bees come back from

Montana we start doing alcohol washes right away to see what the levels are. They come back broodless, because it's cold there, so we treat one more time when they are broodless as the mites have nowhere to hide. The thing is we are sometimes finding that bees from elsewhere are drifting into our colonies and bringing mites with them.

Revenue splits

Steve: Could you say on average how your revenue is split between honey, queens packages, pollination and so on – do you mind sharing that?

Ray: I'd say queens and packages is probably half of our revenue, and pollination...we only do almonds now, not other crops anymore because there are way too many chemicals, way too many

Steve: Almonds are ok?

Ray: Well, we have to do them because we need the revenue. A typical revenue for almond pollination is $180 to $210 per hive. If you do sunflowers, watermelon, cantelopes you're looking at $25–$40 per hive. There's some speciality crops, like there's a type of apple where they have a cluster of flowers and they only want one pollinated, literally. There's six or seven blossoms and one opens up – they want that one pollinated, and the bees might be there for 2 days, then you've got to get them out because they start spraying.

Steve: It sounds like almonds are better – much more money.

Ray: Yeah, but what you've got to deal with there is your bees being around everybody else's bees. It's from Red Bluff all the way down to Bakersfield, the whole valley. For us it's queens and package bees, then pollination and then honey. For honey you never know if you are going to make a profit for sure, so that's a tough one for us to budget for.

Steve: Presumably if you have a bad honey year you at least break even?

Ray: It depends on your inputs...

Steve: Do you ever get years where you make virtually no honey?

Ray: Well last year we made 138 barrels on 10,000 hives. This year we made 1,400 barrels. We moved the bees out of California to get away from the intense agriculture, all of the spraying, all summer, and they went to the high prairies where they have plenty of pollen, even in poor years. The bees come home healthy which means that as long as we do our management right, as long as you have pollen, then you have healthy bees. They can withstand so much more when they are well nourished; it's the same as people. With poor nutrition you get a weaker immune system and you are more susceptible to disease.

Finding the apiaries and breeding queens for performance

Steve: These sites in Montana – how do you get them?

Ray: Well, you've got Montana, North Dakota and South Dakota, Idaho, Michegan – all these places out there produce honey to some degree. If we want locations in Montana we buy a beekeeper out.

Steve: Right.

Ray: They are registered locations within the State of Montana, so nobody can move bees within a 3 mile radius of you. Montana is not a big honey producing area, not like North Dakota, but North Dakota doesn't have a bee law and guys will sit on top of you. Of course, honey depends on getting enough rain.

Steve: I read somewhere that you operate out of California, Montana, Hawaii and Canada?

Ray: Well we work with a gentleman named Albert Robertson. He was commissioned to make a bee suitable for Canada. Albert has been keeping bees for a long time and he's a Professor at the University of Saskatchewan. You can read about it on our website, the Saskatraz Project.

He's developed bees through recurrent natural selection. They need to be economically sustainable and people talk about varroa resistance and tolerance. It all depends on how you define it, but I don't like the use of the word "resistance" because people have over used it to sell things. We don't actually have a bee that is varroa resistant but we have bees that are more tolerant of environmental factors.

If you select for economic sustainability, OK, they've got to be great honey producers, then right there you are already selecting for other things. They are collecting pollen and making propolis. These Saskatraz bees make a lot of propolis and we know that it has medicinal value in the hive. Most of the bees in our hives are Saskatraz. We don't select for colour, we select for performance. The backbone of everything in our hives is Carniolan and Italian stock; we've got a little bit of Russian and some Buckfast in there out of Canada.

Albert got that pool together and just started selecting, but kept track of things. What he has done…he'll never get his money, his time, any of that back, it's just a love for the bees. He's super smart and practical. Anyway these bees have done really well for us, and we're seeing that not only are they good honey producers but one of the best things is their weight and the number of bees going into the winter, and almost the same number of bees in January and February. They are very efficient and we're seeing overall a generally healthier bee. It's selection, we've got a lot of hybrid vigour too.

[bizarrely at this point Ray received a call from Albert Robertson, but said he'd call him back later]

Steve: So he's in Canada, and that's where you are getting some of your genetics?

Ray: Yeah, the Saskatraz genetics. This year I think he sent us about 150 breeders. He comes down in February and we go through the queens. It takes us about two weeks. We look at the queen retinue, the temperament, brood patterns, pollen, how heavy or light they are, we do a mite count and hygienic test on them. Basically we grade the queens. We also ask our customers to track honey production, over-wintering and all that kind of stuff, so we get as much feedback as we can.

Methods used for queen rearing

Steve: I saw somewhere that you use a Cloak Board for queen rearing in Hawaii, is that right?

Ray: For the queen right rearing? Yes, we're doing that over there in Hawaii. We've got a board with an excluder on it so the bees can move across but not the queen. We raise the young bees down below then we smoke them up and Cloak them, so they can't go back down, but those bees haven't had a flight yet. When we graft we want to put them into a lot of nurse bees that are really well fed so we feed the nurse bees up for 3 or 4 days and then let them travel again. We'll graft into there several times and we've got anywhere from 150–200 cell builders. At about 9–10 days we bring the cells in, so they are also the finisher; we don't have starter and finisher colonies, and the cells go into an incubator. They are all kept at the same temperature so they emerge at the same time.

Steve: You have some big incubators then?

Ray: Yes, they each hold about 2,000–3,000 cells. What we do in California is use queenless cell builders.

Steve: Is there any reason for the different methods?

Ray: Availability of bees and labour. We learned the way that they do it over there in Hawaii and we followed what works best for them. Here in California we have hives that want to swarm, that are growing because we've just come out of the almonds and they are supercharged with lots of feed. We go out and collect frames of brood and pollen from hives that we left in the foothills away from the almonds and away from sprays, and use them in our cell builders. We've learned that it's better not to use frames from hives brought back from almonds. It's the sub-lethal effects of the sprays; we can't necessarily see them but they are what I'm concerned about.

We set the queenless cell builders up including a frame of capped sealed brood, mostly to hold the bees, but there's a rainbow of fresh pollen and nectar. We give them a frame of honey and then shake in 6lbs of young nurse bees that came off the brood frames from hives out in the field into every one. The next day at noon we start grafting. We graft in one frame with three bars on it, and each bar has 15 cell caps, so there is a possibility of 45 queen cells. Depending on the time of the year and who's grafting we average around 40 cells per frame, but we budget for 38. If conditions are right we'll do more, because you have to maximise the opportunity when it's there.

If we have too many cells and can't use them we'll call another queen breeder – "do you need cells?"

Anyway, those are also our finishers. We take them off at 9 days and at 11 days we make up our nucs. We use the Mann Lake mini styro nucs and we run about 60,000 of those. We get 3 nucs per pound of bees so it takes about 20,000lbs of bees to stock those nucs. We make them up every day. Our goal is to make up all the nucs before we cage the first queen. We've had a drought here in California for the last ten years; we've had perfect mating weather, which is awesome. The drawback is that by mid May we're dry.

So, going off subject, 50% of our problem is simple lack of forage for the honeybees. They are definitely canaries in the coal mine.

Anyway, we pull the cells then make up the nucs on the 11th day. We shake the bees the day before then all we do is take a spam can on a stick – it's pretty high tech – and we dip the bees into these mini nucs. We'll start at about 3am and finish 9am to make up 5,000 nucs, then the next morning we do the same, and the next morning. Then after the third day we are taking nucs out (for queen mating) and also making up more nucs at the same time.

We just rotate these four buildings. They are cooled with swamp coolers, which keeps it at about 62°F. We leave them in the shop because it's clean, with no smells, fresh cool air, quiet and no light. They are calm; the queen hatches there and we give them one frame of foundation to draw out while they are in there, because they are young bees and they like to work. First of all they want to get that queen out and get her mated, and then they are drawing foundation. We take them out at night to the mating areas.

There's a point where we get really busy making up nucs and taking them out. Our farthest nuc yard is about 20 miles away. These are places where we have our drone mothers and we'll run anywhere from 2,500 to 3,000 nucs in each yard, but we also have the support of maybe 500 colonies within flying distance, for the drones.

Steve: Are those specially selected?

Ray: All of our bees are good. Why would we use anything in our hives that isn't the best? Also, we want the genetic diversity. Sue Coby has always said, "don't breed for colour, breed for performance" and what you want is hybrid vigour. That's why I think our Saskatraz stock is doing so well, Albert said this would happen. We've got stock that's pretty much not ours then we are crossing it with our stock. We are getting the traits built up over time but we are also getting a lot of hybrid vigour. I'll also get breeders from Kevin Ward and Leonard Pankratz just to keep it mixed up.

Steve: What next, after the queens are mated?

Ray: Then we cage them.

On being a family business

As we've grown and scaled up, and our kids came into the business we asked ourselves, "how can we be entrepreneurs and still be a family?" My wife's side of the family is very creative. We have a lot of ideas and a lot of individual personalities in our family. Their concern was about still being a family ten years from now, and not blow up. We hired a company called Leadership One to help us with succession planning and moving from being entrepreneurs to something more professional. It's been about structure and our organisation chart and accountability. We wanted to improve our IT because we want to use metrics in our business. The main thing is that quality is number one for us; the health of our bees and the health of our employees.

The first thing we had to do was figure out how to run a company at that level. We hired a company called Leadership One to help with this.

[Interview continually interrupted by phone calls – just got home from Ohio last night & leaves for Idaho tomorrow morning]

Ray: Their main goal was helping us to structure our business; what's our mission statement, what is the culture that we want to create here, because our bees are only as good as the people that work for us, right? Our kids, everybody that works here, have a performance agreement. We want to make this a place where you can have a career, raise a family and send kids to college.

Anyway, we now have managers of departments, but we're trying to keep it as flat as we can so that we don't have too many layers.

Steve: Do all of your children work in the business?

Ray: All three of them; Haley, Ryan and Josh. They work in different areas and they all have their different strengths. Ryan and Josh both work in the bees and in fact have some of their own bees, which has been really good. They go out and work with the crews. We want everybody to respect everybody but you've got to have accountability.

Steve: Do you have performance targets or whatever...

Ray: Yeah, and they're reasonable ones like "show up to work on time."

Steve: That sounds reasonable

Ray: I does, doesn't it!? Sometimes it's like, "what part of this don't you understand?" I think we're pretty fortunate. We have a lot of depth here. We have a lot of people who are capable of doing a lot of things. Eric Foster who manages the Mann Lake store in Woodland, says that he can probably see this better than I can. He works closely with us and sees how our people interact with him from different parts of the business, like customer services and accounting. That's good to know.

Brexit, equality and developing the labour force

Steve: We've got this Brexit thing coming up in the UK soon...

Ray: Oh boy

Steve: What I've seen in agriculture is that very few Brits will do the kind of work that needs to be done at the lower levels, like out in the fields, so they come in from Eastern Europe. They work really hard and are OK with the money. People are worried that Brexit will stop that.

Ray: Yeah, somebody mentioned that the other day. I think it was Bret Adee, about Brexit.

Steve: It's funny isn't it? I don't know what percentage of local California people work in your company versus people from other countries...

Ray: Probably about 50/50

Steve: You're providing career progression and security – the things that will encourage all people from all backgrounds to work for you – that's the idea isn't it?

Ray: We're looking at H-2A now

Steve: Is that some kind of quality standard?

Ray: No it's about legally bringing in foreign labour. These people come in from Mexico or Guatemala or Honduras or South Africa, they are making good money, and beekeepers say they couldn't do it without them. They are the nicest people, they are humble and they work their butts off. They go home heroes! They make good money, exactly what you were saying.

Steve: Yeah. Some people wrongly say that they are stealing the jobs of the local people but the local people don't want to do it.

Ray: They don't want to work! We see that here. Not quite so much here, more out in the Midwest. We're also dealing with the take-off of the fracking industry. They are starting entry level work at $27 an hour, you know? A lot of those guys that did have locals working for them lost them to that. It's about short-term gains, not long term.

Hard times and disasters

Steve: Right. What about any particular memories of disasters or tragedies that you have had on your journey?

Ray: Well, there was the El Nino in 1983

Steve: Oh yes, with them getting washed down the river

Ray: That was difficult, and at that time bees in California weren't even considered agriculture. Now there are small business loans and disaster relief funds. It was really difficult when the varroa mite showed up. The learning curve was very steep and everybody's first impulse is to quarantine everything, which does nothing for anybody. If you're going to get something, whether it's tracheal mite, varroa mite or small hive beetle, if the environment is conducive to it then it's going to happen. In Hawaii I thought the larva rocks would stop the small hive beetles from multiplying, but they can multiply on the bottom board in a little bit of debris. I can't think of too many things that were disasters though.

Steve: I was a bit worried about these fires that I keep hearing about on the news, but they are not near you are they?

Ray: Yeah they are 50 miles over here, 30 miles over there but we keep our bee yards clean. The liability to fire is high, it can put you out of business, so we do safety training all the time so that we don't cause a fire. We are always worried about somebody else burning our bees down but our yards are spread all over the valley, it's not like they are all sitting in one spot. There have been beekeepers who have lost quite a few bees. Over on the Big Island David Thomas lost a lot of bees. The Florida beekeepers have suffered badly with hurricanes, and probably Louisiana. We get bears and things like that, but nothing that goes as far as a catastrophe, thank God.

Being a good beekeeper - some tips on the craft

Steve: A lot of people who are going to be reading my book, hopefully, are going to be hobby beekeepers, as well as the pros. I think it's interesting that a bee hive is a bee hive, isn't it, a colony of bees is a colony of bees so you still have to inspect it every 7–10 days at certain times...or do you?

Ray: Well, once you learn about the bees...First of all, you need to know your environment – where the bees are at. If you've got good flows and things there's times when they will build up so rapidly that they'll swarm on you. That can be a problem if they don't raise their own queen, but it's a potential time to make growth. If you split it before it swarms you go from one to two hives; that's 100% increase, and then you try to keep both alive. It depends on the time of the year when you need to inspect them. It's fun to inspect them when they are doing well.

Coming out of the winter you need to inspect the bees to assess the strength of each colony, and their overall health. You need to do an alcohol wash – learn how to do that and have a good treatment plan for the mites. You should treat the bees; we are beekeepers – keepers of the bees, OK. There's a notion out there that we need to let the bees go and do their own thing, but if that's what you think, stay out of the bees. Don't do the bees because you're keeping ten hives and if you let them all go and just try to keep survivors then you are putting a crazy amount of mites into the system. The bees can't fight that because the mite numbers are too high. Let the bees in nature take care of that but if you are going to keep bees, take care of them. Randy Oliver will tell you the same thing, and Michael Palmer said the same the other day; he almost screamed it out. You know, treat your **** bees!

You need to know about nutrition and mite levels.

Steve: How is this knowledge of the craft of beekeeping passed on? You have been doing it forever, and your Dad kept bees, so do you have training programs for people new to your business?

Ray: We try to train them out in the field. We are working on that now. We have our regional areas, where a group of 5 or 6 guys work one region, then there's another region here, and another here and so on. That group of guys takes care of their region until we go into the almonds, so they get to see those bees again and again, and they get to see what they are doing right and what they are doing wrong. The way we worked before you might not see those bees again for six months, depending on the rotation, but now we want the guys working within a region so they can see their bees progress.

We can then incentivise those guys for doing a good job. If they've got 2,000 hives and they only lost 50 and the rest went to the almonds, "Good job guys!" What you have to be carful of is that this region over here may be great for flowers, pollen and everything, whereas this one might be up higher in the foothills and need more supplemental feeds, so there's more that can go wrong up there.

I got a bit lost there – what were we talking about?

Steve: The passing on of knowledge.

Ray: Oh yes. The thing that I've learned is you can tell people what to do but until you experience it with bees, I don't know why it is, but you just have to live it. There are so many variables and you never quit learning. If anybody tells you that they have it all figured out they are so full of baloney. They haven't figured it out, they are probably going to collapse at some point. The more reactive you are the more chance there is of problems. You've got to be proactive on certain things. I'm already thinking about next year at this time. We've got good clover that came up and germinated in the springtime and this fall, because we had good moisture out in Montana, so we could have a bumper crop.

A bumper crop can be as bad as a terrible crop, maybe even worse, because instead of making 3–4 supers of honey you make 8 or 9 supers of honey, which means you need more labour. It takes you longer to get the honey off the hives and then extract it, so you are treating for mites later. I know guys this year who had good crops, they are still taking honey off and they should have treated a month ago. They're gonna collapse, I know it. It hasn't happened yet but it's going to. I'm already thinking about how I can deal with this, because if you just react to it you are too late.

Steve: That's interesting. For a lot of beekeepers in the UK their main crop is from the heather, which is a late crop, so they are only just treating now. We are just about into winter now.

Ray: Yeah, I know guys that have had two crops. You get an Indian summer, they've already pulled their first crop down, it rained, and it comes back. The problem with that is the bees don't have room to store the nectar so as the brood hatches they start filling in all those cells, so you don't have space for the queen to lay the bees she should be laying – the winter bees. You end up with a dwindled down colony in the spring, and then you throw in viruses if you don't treat on time, so you have to have a plan using formic acid or oxalic acid so you don't get residues in the comb.

What's more important, the extra honey or keeping the bees healthy? We are very mobile. We can move 1,000 hives in one night. You can pick them up and move them somewhere else where they are not doing so damned well! If you leave them there, what we sometimes do in that case is that we bring in hives from another area where there is no flow and saturate the area so they don't get too much in any one brood box.

Steve: That makes sense. If someone finds queen cells what do they do?

Ray: We just knock them out. We have queens on our truck. Every truck that goes out, every day, no matter what they are doing, has a bank with queens in it, because we raise queens. We replace queens as needed. If it's queenless we steal brood from another hive and add a new queen. If it's got queen cells, she's still got to go out and get mated, and if she doesn't then the hive's no good. We give them the queen that we want them to have, and because she's mated we don't get so much of a brood break. Once they start raising cells they kind of have a mind of their own. "We don't need a queen, we've got these cells." It only takes a matter of minutes before they know what they don't have.

Inventions and brilliant ideas

Steve: Have you invented anything? Any brilliant ideas to solve problems?

Ray: Not really. We've done some things to make jobs easier, like filling the nucs. Each one has a little feeder inside and we would fill them with syrup using a watering can, one at a time. We'd been doing that for 60 years, but every one drips and you can track it through the shop. It's a mess. We called Mann Lake and got piston fillers. We got a 225 gallon tank and these 4 piston fillers and an assembly line, so now instead of taking 10 hours to do that with 8 girls, 4 or 5 of them can do it in 2 or 3 hours. It's about improving our efficiency – not so much an invention as constant improvements.

Steve: Things like making up frames or boxes – do you have people making them or just buy it in?

Ray: We buy it in all assembled now. We would rather have people in the bees than building equipment, and we don't have much of a winter. Look at our area – we have bees flying nearly every day, there are not many days off.

Steve: Did you talk to Mike Palmer much? He has bad winters.

Ray: Yeah, it's like Canadian winters.

Steve: It's so different isn't it? The same country but...

Ray: Yeah. It's really important for him, just like the Canadians, to have a good queen and good pollen stores and good nectar stores, and a strong cluster to get through the winter.

Controversy - imports and environmental concerns

Steve: In the UK these are controversial issues...We tend not to have too many packages imported but we import a lot of queens. Many people think that is a bad thing because of disease risk, but the bee farmers need them, and if you haven't got them you need to get them from wherever you can.

Ray: People say you should only get queens from your area because they are best adapted to your conditions. We've sent queens to Canada and they make a huge crop and over winter well – to me it's always about who drives that car. If you give a Porsche to me and one to Mario Andretti, who's going to get the best out of that car? It's the person with the highest skills and experience. We can get the most out of our bees because of how we take care of them. Italians are going to need a bit more feed and they'll have a bigger colony, and you need to know that if it warms up quick in the springtime they're going to start brooding sooner. You just have to understand the bees. Carniolans are more forgiving, more conservative.

Steve: You've already said you have concerns about lack of forage – how has that changed over the years?

Ray: It's a big thing in California. There used to be a lot of small dairies, and all of them had clover. This used to be a big area for clover seed and we'd get paid to pollinate clover fields. The almonds, walnuts, pistachios and the tree crops going in have changed all that. They are looking at ways to mow less; instead of mowing every 2 weeks they just spray with Roundup and you've got bare orchard floors. The fence lines and roads are clean – the State sprays all the weeds along the roads to reduce fire risk, so it's a snowball effect. However, we are now getting a lot of orchard owners that are planting back cover crops. They understand that not only do bees pollinate a third of what we eat, they also pollinate the clover and alfalfa that feeds the cattle. In the United States we have cheap, available food, and bees are one of the reasons for that.

Steve: Yeah, I was in a Safeway over here recently, and it just has everything you could possibly want on the shelves. If I lived here I would just eat all the time and explode. Anyway, another big deal in Europe is the ban on neonicotinoids. Obviously, if you are a beekeeper you are not a fan of insecticides, but many commercial beekeepers I have spoken to didn't have a problem with the neonics. They couldn't really see any detriment in their hives. They are more worried that if we ban neonics maybe we go back to some of the nastier stuff.

Ray: Organophosphates, yeah.

Steve: Oilseed rape is quite big in the UK and it was treated by a seed coating.

Ray: I don't know too much about that. I think it's part of the problem, and there's timing issues, but it feels to me maybe things have to get worse before they get better. Politically everything is so far left or so far right, it's really disturbing. I don't know enough to really comment but I hear from some people that it is a problem and some not. The main thing is our environment. If the bees have proper nutrition they can get through a lot of these things.

Ventilation and insulation

Steve: What about ventilation and insulation in a hive. In our country you insulate the roof all year round and we only use a bottom entrance. We don't use top entrances. In America it's the norm to have top entrances isn't it?

Ray: Everybody's different. We have a hole in the top of our lid which we use for feeding syrup using an inverted can. We get more stimulation with a slower dripping effect from the can rather than a more rapid feeder. In California this year we got down to 10%–12% humidity. We get some in the winter but the bees regulate themselves pretty well.

Steve: You don't deliberately have upper entrances? You just have the feed hole at the top?

Ray: Yes. In fact, with all the bees that we have around here it's probably best to have fewer entrances.

Varroa treatments

Steve: OK, what about varroa treatments – what do you use?

Ray: We've used oxalic, we've used formic acid with variable success rates. We've used Apivar, which is Amitraz…

Steve: That's the main one isn't it?

Ray: Yeah. Amitraz has been around here for forty years as far as bees go. It works on tracheal mites and varroa mites and it breaks down instantly. It's more of a flash treatment, other than the strips. The strips are made of a dense plastic impregnated with Amitraz, which migrates to the surface over time. Some people make their own concoctions. You've got Apiguard, which is thymol; that's very temperature dependant.

Steve: It's probably too hot here isn't it?

Ray: Not in the springtime. Even right now, if you could stay in the low 80s – middle 70s it would work OK, but again, it's not completely effective. Some mites remain, and if the bees can fly and you have no forage then they'll go into other hives and you've got re-infestation. There are just so many variables. On a small scale I could probably keep mites at bay with thymol. We use an essential oil patty that we keep in the hive most of the year. It has thymol, wintergreen, spearmint, eucalyptus, camphor oil and lemongrass oil which are anti bacterial and beneficial.

Steve: Have you have seen that it definitely helps?

Ray: Yes.

Steve: What about feeding?

Ray: As needed. We get it from Mann Lake; it's got 77% solids, it's a high fructose corn syrup/sucrose blend. There is some sucrose in there which the bees have to invert to break down, which causes stimulation.

Steve: Do you weigh the hives?

Ray: No. We don't have that over-wintering. If we did go to over-wintering in buildings we would want to make sure the hives weigh a certain amount; that's the future for us – indoor wintering.

On growing the business and looking after the bees

Steve: One of your plans for the future that you have already mentioned is going into processing and packing, wasn't it?

Ray: Yeah, producer-packer.

Steve: In other words you will be shipping out jars of honey with labels on. The guy I saw in New Zealand who owns Airborne Honey actually stopped beekeeping and completely specialised in packing and processing. It was all about how to differentiate your products from everyone else, having a good reputation and mono-floral honeys and so on. I guess you are about to start on that are you?

Ray: We have been doing that. Chico Honey Company started 4–5 years ago. Our kids started this label on their own. They started bottling honey by hand, which helped to get them through school. That's going to stay as a regional label. Once you start to get bigger there's things like compliance, so OHB is helping out with that stuff. Being a producer, we know our honey, and we have the family story, so it's a good fit for us.

Steve: What would be the main types of honey that you sell?

Ray: Well, we have our Hawaiian honey of course. We want to pack and sell our honey over there because it's all tourists over there. We have our Montana sweet clover, and we have a general Montana wildflower honey. Yellow sweet clover, when it does bloom is just about as close to a single source honey as you can get. It doesn't want to granulate; it's like yellow star thistle used to be, or knapweed, or sage. Then our California honey is mostly mountain wildflower. We just don't make much in the Valley here anymore because there's too many bees.

We don't go to the citrus. We don't like putting our bees anywhere where they might get overly stressed, just to make honey. If we want to sell some other varieties we'll probably get it from some of our queen customers who are commercial beekeepers. We have good honey sources and we know everybody, or a lot of people.

Steve: How about your legacy? What do you want your legacy to be?

Ray: Wow, legacy?!

Steve: I know, you're too young for that right?

Ray: If anything, that we were a great place to work and we took care of the bees.

Steve: That's great! If you could have any wish granted, just to do with beekeeping, what would it be?

Ray: I'd have to say that we improved our environment for the bees, because the effect of that would be beneficial to our kids and grandkids. If we start taking care of our environment better, I don't think you could even write down all of the things that would improve as a result of that. It would improve the wildlife, the health of our children and grandchildren. Bees are definitely affected by the environment and are a good indicator of what we're doing to ourselves, I think.

Predictions

Bees are incredible, so resilient, all we can do is try to give them the right conditions. I think genetics will get better, they'll figure out something. In fact there was an engineer from Ohio who approached me at the recent Bee Culture event, and I can't remember what he was referring to, but something revolutionary is going to happen in the next 20 years; we just don't know what that is right now. He was talking about gene coding and stuff.

Steve: Yeah, that worries me. You were talking earlier about keeping diversity but I think they have now got the full genome of a honey bee, so they can possibly artificially create bees that don't occur in nature – man made bees that might be perfectly designed for something we need.

Ray: Everybody will tell you that when you select for a certain trait, just like when Tabor was selecting for foul brood resistance, you know, he lost honey production. Every time you select for something you have to be careful. That's why Albert Robertson talks about economic sustainability; that's honey production, that's gentleness, over wintering abilities and things like that. We can make really good improvements but we need to be concerned enough that we don't screw it all up.

I think that the future of beekeeping is bright and we are in our infancy.

Future plans

Steve: For your company are you continuing to grow, or just diversify, what's the plan?

Ray: We are growing but only at the rate where we can keep being the best, or one of the best. I don't want to say we are better than everybody else; I can't imagine anybody not trying to raise the best product that they can raise. We might have some better environments, and maybe do some things that are a little more labour intensive to ensure quality, but I feel like everybody does a good job, or can. We want to do agri-tourism which is more about educating people, and then we've got our store, and we're helping set up a bee museum in Orland – so we'll have microscopes and we'll have this one building that you'll walk into and it'll be like being inside a bee hive. We want to make mead...

Steve: Is that becoming popular again, mead?

Ray: I don't know if it ever was popular. When you ask people if they've ever tasted mead they say, "meat?" When you watch Game of Thrones they are drinking mead, Friar Tuck in Robin Hood had an apiary for the mead. There is so much that you can do with mead; it can range from 3% to 27% alcohol, you can distil it – we've got an 80 proof. We may not even do mead because the distilled stuff is just incredible. It's a high-end product in the $60–$80 range for a bottle. It takes about 5 bottles of mead to get one bottle of the distilled product.

Steve: Are you one of the bigger employers in Orland?

Ray: Oh, I don't really know.

Steve: Are you well known in the area?

Ray: Well, they are starting to get to know about us. The City took on "Queen Bee Capital of North America" because over half of the queens produced in the US are produced in Northern California and we are a big slug of it. The City wanted an identity so we are a "Bee City" which involves trying to help the bees, the environment and things like that. We just had Oktobeefest at our place where 200 people showed up, and for our hobby day we had over 3,000 people here. Only 700–800 were people who wanted bees, the rest wanted to see what bees were about and taste honey.

By the way, please call me anytime if you have any more questions or want to check stuff with me. I hope you do. We had National Geographic chasing us for a couple of years trying to do something, but they didn't want the upside, they wanted the downside…

Steve: Yes, there's a lot of that, there's an agenda out there which is all negative: "the bees are all dying and we're doomed," – is that what you mean?

Ray: Yeah. I think we are way better beekeepers than we were 30 years ago.

Steve: They're not all dying out anyway, are they?

Ray: No. No, in fact what we are doing is slowly creating an environment in which they do better. It's happening slowly. For a long time people would ask me, "How do you make any money keeping bees?" We pollinate, we raise queens – people just didn't know. We have over 160 employees between here and Hawaii. We can make more money keeping things small but we enjoy doing this. One of the hardest parts of the business is the employees but we are trying to put things in place. We've got really good employees but every so often you get a few problems.

Steve: I'm glad, having seen your mindset and your values; I'd rather you got big than someone else.

Ray: Yeah, I understand what you are saying.

It gets easier. The more infrastructure you have the easier it is to grow. In Montana we are taking on another 3,500 hives and some more locations. We are trying to get all of our bees out of California. I'll probably sell some of the bees because we've got as many as we want to run, but we are trying to get the bees into an environment that is healthy for them. It's going to take a few more employees and a better structure out there.

Recently I was working more in the business than on the business for a while, just three months ago. I worked every day for two months in Montana, because it's not like work to me. I took one day off and flew with a friend to a cabin in Idaho on the Snake River, but I was out in the field the rest of the time helping, because we were short of employees...

Steve: And you love it!

Ray: Oh yeah, it's not like work. That's what people always say, they say, "we'd rather see you go fishing," but if I'm splitting a hive or looking for queens on a Sunday, that's actually a lot of fun. It's nice where I'm at – that's just the same as catching a trout or shooting a big buck, you know?

Steve: Do you get much family time or down time? I know you said that you have family meetings, so that you have time to be a family rather than work colleagues.

Ray: When we hired Leadership One, one of the first things was our mission statement and the culture that we want to set for our employees and our business. The second thing was, "How do you work as a family, and respect each other, how do you talk to each other?" Even this morning we got loud with each other because we have different ideas, but we have to be able to listen to each other and respect each other. We know way too many family businesses that just don't make it.

A small family business came to us because after their kids came back to work in the family business they were about ready to explode after a few months. "I'm doing more than he's doing, he's not doing what he said he was going to do, he doesn't talk to me right."

Our first meeting that we had as a family involved pencils flying across the room and some crying – it's taken us a few years to get to where we are now. Things are getting a lot better now.

Steve: Great, many thanks for your time Ray.

Wintering queens in quad box

Peter Little, Exmoor, UK

'That's what beekeeping is; you've got to have good intuition, a good feeling and be studying and watching bees, you've got to be in touch with them. You can't be remote from them.'

Introduction to Peter

I knew that Peter Little was a nice guy, having had a few telephone conversations with him at various times, but I had no idea until I met him just how highly skilled he is in so many technical areas. He is all about being self-sufficient with all aspects of his beekeeping and queen breeding. He does not need to employ staff because he has five able sons and his wife, Sandra, who all play their parts in the endeavour. They use a sawmill to cut Western Red Cedar for use in hive building, their hives are all self-built, as are the frames that go in them, and the wax foundation is made on site.

However, it is in the area of breeding honey bee queens of the highest quality that Peter truly excels. He uses instrumental insemination as well as two isolated mating stations (one being the famous site on Dartmoor used by Brother Adam back in the day), to have control over which drone genes get joined with the genes of selected virgin queens. There are many people selling queens who are not so much queen breeders as queen rearers. They buy a few breeder queens and then graft countless larvae from these queens, which are put in cell builders and generally open mated, to produce large numbers of good queens for sale. Queen breeding is about controlling both the drone and queen lines, and it's about having an instinct for which combinations are likely to do well.

My mind was blown when Peter started to tell me about how he makes a lot of his equipment rather than buy it, partly to save money, but partly for the love of doing it. For example, he makes his insemination tips from soda-lime glass capillaries. To me, the idea of sticking a tiny sperm filled glass tube into the nether regions of a prized queen bee is terrifying, and I'm sure the queen would concur, but Peter has been doing this for decades. I was there to interview a beekeeper and found myself with some kind of genius bee surgeon. When I visited several months later, Peter proudly showed me his latest project; a self-built honey drying machine. This will reduce the water content of honey that is harvested before the bees have capped the cells.

My wife Elaine accompanied me on my trip down to Somerset. It was February 2018, and daytime temperatures rose to about 5°C. Beekeepers are much easier to talk to when their bees are not flying! We stayed in a bed and breakfast establishment in Porlock, which is near to the quaint villages of Bossington and Allerford. Elaine sat with me as I blundered my way through the interview. To my surprise, she seemed quite interested and even asked the odd question. Sandra sat there too, pitching in with occasional gems like, "he knows his bees like a farmer knows his sheep." Mostly, as with many of my interviewees, my job was simply to sit back and listen. When somebody as experienced and knowledgable as Peter Little is talking, it is wise to allow them to unload their stories, no matter how far they deviate from the question asked. The tale of the shrew explosion was one such delightful detour.

Peter has use of the famous Sherberton mating station on Dartmoor because he is friendly with the landowner, Anton Coaker. This is the historic site previously used by Brother Adam for mating and overwintering his Buckfast queens. The subject of the famous beekeeping monk naturally arose during our chat. I think it is sad that after all that Adam gave to beekeeping, there is little at Buckfast Abbey to suggest he ever existed. Perhaps they wish to focus more on heavenly matters, and perhaps they are right. Honey farmers are not likely to be balanced people; they are more often than not profoundly obsessed with their bees and have little time for other things.

Talking with Peter

As we began...

[Peter was talking about a recent order he had fulfilled for bee hives for a customer moving to Bermuda, and I hadn't yet got my recorder on]

Steve: I'll just turn this dictation machine on…in case you say something interesting [laughs]

Peter: We sent a lot of equipment out to Bermuda. It was western red cedar hives, all with frames, all fitted with foundation. Sandra made 400 frames to go in those hives.

Steve: I wondered if they even had honey bees out there

Peter: They do. The hives have arrived now and soon they'll have bees.

Where I keep my bees...

Steve: About your bees – do you have a winter place for them, and then move them in summer?

Peter: Well we've got permanent apiaries, and in some of them there can be 80 or 90 hives there in the winter, and then in the springtime they get thinned out to the summer apiaries. We've got dozens of them out on the hills. Out there it goes up to one thousand and seven hundred feet. We are down in a micro climate here. You wouldn't think so, because its cold here now, but we are three weeks at least in advance of everything out on the hills. So, the bees build up down here. They get a honey flow and a good build up in the spring, and by the time they've done that things out over the top are just starting, so it's as if they are getting a second spring. That's when we put them in smaller apiaries, usually between twelve to eighteen hives to an apiary.

Our apiaries are surrounded by heather but there's a wide range of forage here. We have a whole lot of sweet chestnut trees, so it's white with sweet chestnut during July and into early August. We've got lime trees, sweet chestnuts, bell heather – all of the apiaries can reach heather – some have only just got to go over a fence, or a couple of hundred yards and they're on the heather. We get a cross over point where we've got to get any honey off that's normal (floral) before they start piling the heather in. They are permanent apiaries because they do well on bramble and other forage before the heather, and then they collect from the heather later on too. They don't need to be moved.

Then we've got specific heather apiaries. Some of the bees get moved after spring to a place where they get rosebay willow herb and clover mainly, if the weather's right, and other forage. Then in July they get moved to specific heather sites, just for heather, and when the heather's finished they get moved back down here to the wintering sites ready to do the cycle again in the spring.

We've got over five hundred hives and we're building up the numbers every year, more so now because of the boys. I would be quite happy to be going down in numbers but because of my sons, who are carrying on with it, we need to build up the hive numbers.

Steve: OK

Business and Operations

Peter: But we haven't got all of our eggs in one basket with bees, because we sell queens and nucs, and bee hives, and honey as well. So, we don't just rely on honey.

Steve: Would you mind giving me a rough breakdown of how your revenue is split between these different income streams? Not the actual revenues, but the proportions?

Peter: I suppose a half would be honey – this is just off the top of my head – and the other half split between bee hives, queens and nucs, probably about equally. I would like the hive numbers to go down to be honest.

Steve: Is it just more work?

Peter: I'd sooner be working with bees than making bee hives for people.

Steve: And queens is a big thing for you, isn't it?

Peter: We sell a lot of queens, and the numbers grow each year.

Steve: I remember when I finally tracked down Murray McGregor, because as you know I've been trying to track you and him down forever...

Peter: Murray's is a much bigger business than ours, obviously. He needs to employ about twelve people I believe, for the summer work, whereas we don't have to employ anybody, because we have our sons. We may take on an apprentice as we grow; we'll see.

On breeding queens and artificial insemination

Steve: Before I met Murray we emailed each other, and I think he told me that you were more technical than him at the actual queen breeding. He doesn't do instrumental insemination.

Peter: Well, I don't know, Jolanta is very good at what she does.

Steve: Yes, she's very smart

Peter: I've chatted to Murray on the telephone several times and I met him down at one of our bee farmers meetings in Devon.

Steve: He was very complimentary. He said if I want to talk about queens then you are the technical one.

Peter: If you mean the breeding part of it then yes. Last season I inseminated 41 queens. You can make good headway with inseminations. I inseminated 41 but not only that; from those inseminated queens I managed to then raise 21 queens from one of them and they were open mated. So that's a breeder queen, daughters from her inseminated, from the inseminated queen grafts taken, and daughters from her raised then open mated, and they're now overwintering in colonies, mainly 6 frame nucs. So, I can tell what those daughters are going to be like from the queens I inseminated. That's called daughter testing.

Steve: Right

Peter: From those daughters I can see how good they are. For a start, overwintering. How much food did they use? Now for this coming season, how productive are they? Are they swarmy? Their temperament? I already know that's good. We don't wear bee suits here.

Steve: Do you wear a veil?

Peter: I don't wear a veil, it's over the back of my head. I might put it on once a week. We were going though overwintered nucs yesterday putting on ekes for feeding and they're flying around but I don't take any notice of them.

They're not aggressive; they're good tempered bees. The boys never wear bee suits but they wear these BB Wear vest type veils. Lewis quite often has his off but Andrew, when he's working on the bees, pretty much always has the veil over. He wears a tee shirt and jeans, not a bee suit. This is what I wear for beekeeping, what I'm wearing now. I go through bee hives just like I am now. Except if it's hot weather.

Steve: What about your beard, do they get caught in that? [Note: Peter had been instructed by his wife Sandra to shave off his beard and get a hair-cut for my visit. Normally he has a beard]

Peter: No, never had that problem Steve. If it's hot I'm in a shirt with the sleeves rolled up. I don't wear tee shirts but the boys do. This is not just inspecting bees, this is shaking bees out for packages, shaking them off frames making up nucs, everything. We do it in shirts and tee shirts. We don't wear bee suits.

Steve: They've got to be gentle bees for you to be able to do that

Peter: Yes. Anyway, going back to the queens, what we're doing is daughter testing. We know then that it's repeatable. I can go back to those inseminated queens and know that I can graft good daughters from them. Not only that, I can also go back to the breeder queen and the drone line and do some more inseminations to produce more of the breeder queens.

Steve: Yes. That's where the instrumental insemination is so good, isn't it?

Peter: We do that and we do isolated mating as well, it's the same thing, but isolated mating is good for larger numbers of queens. It's not 100% guaranteed like insemination is. It only takes one stray drone to get into your isolated mating to mess things up, but it's good for bigger numbers. It's slow progress. You can only really do one drone line in a season. You can raise several queens from several breeders but you are only mating them to one drone line.

With insemination what we do is put drone combs into various colonies we wish to raise drones from. So, we'll put a drone comb bang in the middle of a brood box, sometimes two, and those can be put in lots of different breeder colonies – ten if you like – which we do. We colour the drones.

Steve: Oh?

Peter: You can only do one drone line in an isolated place because you haven't got time in a season to be putting those drone providers there, the sister queens, moving them away again to put another drone line there – it's too intensive, you wouldn't be doing it in one season. So, you make slow progress.

Steve: OK

Peter: Whereas with the insemination, by putting lots of drone combs in different colonies, they don't have to be isolated – they can be right next to each other in the same apiary…when the drone combs are sealed we'll take that comb out and shake the bees off, then we'll put it in a colony above a queen excluder. There are no other drones up there, just workers, and the baby drones emerge from the comb. I have the dates all written down so I know when they are going to be sealed and when the baby drones are going to emerge. Then we use 8mm chisel tip uni posca marker pens, we have all colours, and we start putting a dot on each drone's thorax. We don't pick them up, they are just baby drones walking on the comb and we put dots on them. We will mark about 500 drones.

Steve: Right

Peter: I'll pass the comb to Lewis and he'll mark some, because you get pretty bored of this, then Andrew will mark some. Then we take the queen excluder off and shake all the drones down into the box because if they are left up there trapped the majority will just die. They are not sexually mature until they are about 14 days old, and even then some of them aren't, so I have found that 21 days from emerging is when they yield the maximum amount of drone semen, which is what we want. You also want them being well looked after by the workers so the best place for them to be is in with the rest of the colony. You want them to be able to fly so they are fit. You want good healthy fit well looked after drones. Never let the colony run low on pollen or stores.

You raise the queens as normal, and the virgin queens are in the mating nucs. They are free running as well. You want them free running before and after insemination. This business of banking virgin queens is not good. After they have been inseminated you want them running around with at least a small patch of brood in the mating nuc, because it helps the drone semen to quickly migrate to the spermatheca.

The headway you can make is because you've got lots of different virgins from different mothers and you can be inseminating them with different drone lines, instead of just one drone line in an isolated mating station. When we want the drones, we go out and if it's a nice day you collect them by resting a queen excluder in front of the hive and just pick them off. If it's not nice weather we just go through the hive. The drones will generally be on the outside of the brood nest. You take the comb out and collect up 60–70 of the drones that are marked. These drones turn up in colonies all over the place. You can go into a colony half a mile away and will see two or three drones that you marked in another apiary, which shows you how they spread.

Anyway, we know what the drones are, what their mother is, what the drone line is, and we know the queen line obviously. Drones will die very quickly. When you collect them up by the time you get them home two thirds of them will be dead. They die of shock. They are very delicate, these drones, they are not at all hardy like workers. Drones just drop dead for no apparent reason. They need attendant worker bees with them, but it's a pain if they are all mixed together, so we use a split cage. It's a box with a little drawer on it which holds a piece of fondant, and it's got a mesh screen on it. There is another section which is hinged onto that, with a hole for a cork in it, and on the face of that its plastic queen excluder. You go and get a cup full of attendant bees and put them in one cage with the fondant and seal that up, so you've got a box basically with a screen like varroa mesh, with all the worker bees in. When drones are with worker bees you can keep them alive until the next day, because they are being fed and looked after and they feel happy and secure.

Steve: OK

Peter: The top piece hinges down and you pick up the drones and you put them in through the cork hole, as many as you want. They are on their own separate side but going against that mesh being fed by the attendants, and being kept warm. Any workers in there can fly out through the queen excluder.

Steve: Did you invent this?

Peter: I didn't invent it as such but I invented my own cage for doing it. It goes back to about 2001–02 when we were inseminating for a whole week. A guy called Tony came down and stayed with me for a week. We were going out doing normal bee work in the morning and later on we had a lot of virgins to inseminate. It was all planned out. The first virgin he inseminated, he pulled her in half. He pulled her sting right out. I thought "bloody hell, this is off to a good start Tony." He said "whoops." I thought…go away [laughs].

Steve: That's one of your prized queens [laughs]

Peter: Anyway, at around midday the drones were flying so I sent Tony out to collect some. We realised that most of these drones were dropping dead. By the time they got back here most had died. Tony works in the NHS [National Health Service] on designing new prothetic limbs and so forth, he's a bit of an inventor and a bodger. Together we made a split tube from two plastic containers, so that we could put worker bees in one side and drones in the other. The idea was to bring drones back with attendants to see if they lived longer.

Steve: I've never heard of that before

Peter: We used this split tube to put attendants in one side and drones in the other. Well, it worked! We brought some drones back and they were still alive the following day, to our absolute amazement. I didn't want to carry this huge pot around so I made a more compact one, which I've used hundreds of times. It just keeps the drones alive. This latest one has a seven Watt terrarium heater in the bottom, because you've got to keep them warm. You've got the attendants in the bottom section and the drones above, and they can fly out around the box and defecate, because they're full of that; it's what drones do. They fly around, but when they feel insecure or want feeding they stay on the mesh being fed by the attendants.

You put your arm through the cloth sleeve to pick a drone out, then you squash its head and thorax, then gently roll the side of the abdomen. As you do that you squeeze it, which causes a partial eversion, then squeeze it not firmly but harder – if you do it too fast it just pops right out and gets on your thumb nail – but you squeeze them until full eversion and their little endo-phallus pops out. It's like a white bed of mucus with a little tiny blob, in a fertile drone, a little blob of creamy buttercup coloured drone semen. You then have to skim that off the top without collecting the mucus, which is all done under the microscope. You need about 8 micro-litres of that, which on my tips are about 12mm.

Steve: That's the really small tube...

Peter: I make the capillary tubes, the tips, myself

Steve: Really?!

Peter: I buy the sodium usually. You can get borosilicate as well but I find it's not very good. I like sodium, the glass capillaries. To buy a tip costs about £35 but for £7 I can buy 200 90ml glass capillary tubes from Germany. I put it in a little vice clamp with a nichrome wire heated ring, and a weight at the bottom and another chuck – this is the proper way of doing it – and it's a tip puller. You heat the ring, the weight at the bottom stretches the glass, it pulls it down to a point. I then use a gauge wire inside, for the right diameter. Where the wire stops is where I snap the glass off. I flame it first, because if you try to file it straight away the glass will split, so I flame it to polish it – just flick it through my lighter a couple of times – or a Bunsen-burner, but I use my lighter, and then I use some twelve hundred wet and dry, with some water to gently file it up and down, then I flame it again and that's an insemination tip.

Steve: Wow

Peter: I've just made a new syringe as well, a high capacity one. Those things cost about £165 each so I just bought the glass for about £3.50 and made it into a high capacity one. That's what you put the saline in, which is salt and water basically; you can buy it but I make all my own. I add the salt and set the pH, then I put it into small bottles that Tony got me many years ago, I can't remember the name of them; they are special bottles with an aluminium cap, and you can autoclave them. They go into blue autoclave bags. I haven't got an autoclave but I use a pressure cooker which does the same job. They go into a pressure cooker and they are in there for about 20 minutes so everything's sterilised. The tips go in there too. Everything gets sterilised, everything that comes into contact with the queen.

I also make my own sting hooks. They are silver, they're even more expensive. I did have a friend over in Ireland that made them. Sue Coby used to use them.

Steve: What, she used to use your ones?

Peter: No, she had them made in the States. They had a watch ruby in them. I had my friend Danny in Ireland send some over, because he had mastered the art of it, and now I've still got the ruby ones but I make my own sting hooks out of a piece of silver. You basically heat it over the gas, flatten it out with a tiny hammer, bore a tiny hole in it so small that you can't see it with your eyes – you have to use a microscope. It has to be able to grip the queen's sting. Even the very point of the finest pin is too big. I use that but I don't go right through with it, the hole is so tiny you wouldn't believe it. I file it with wet or dry, shape it – it takes me longer to explain than to do it.

Steve: That's in silver?

Peter: Yes, it takes about 15 minutes to make a sting hook.

Steve: You've got to have a steady hand or something. What's the ruby for?

Peter: The rubies are in the ruby sting hook. Instead of having a hole bored through silver or steel, there is a bigger hole made into which the ruby is inserted, because the ruby is a watch makers ruby, and they have already got a tiny hole in them.

Steve: Oh, I see.

Peter: They are bearings for a watch, but the hole in them is on the borderline of being too large. Nine times out of ten it will grip the sting okay.

Steve: The queen's sting is not barbed is it?

Peter: No. Once the queen is in the holder she's anaesthetised with CO2, and you use a ventral hook to open up one side of her vent, and on the other one you have to thread the sting through the eye of a needle basically, then the angle of the hook grips it so you can pull the sting upwards and outwards, so that you can insert the insemination needle. But a lot of people use forceps, and that's what Tony was using, and that's how he managed to pull a queen in two. It's like a little pair of forceps, and you grab the sting and pull it up and out, like a tiny pair of tweezers. I don't really like forceps. I don't like flattening the queen's sting. Although the majority of people do that I don't like that method. I think the sting hook is a gentler method.

Steve: And the sting has to be pulled aside for it to work?

Peter: Yes, you have to pull the sting upwards and across. That reveals the oviduct.

'There's a total difference between being a queen producer and a queen breeder.'

Steve: Where did you learn all this then?

Peter: Basically, I taught myself, using the internet, and talking to people like Dr Joseph Latshaw at Ohio Queen Breeders. He manufactures equipment but years ago I used to spend hours chatting to him. Sue Coby, who's the best in the world; I've had lots of help from Sue Coby in America as well.

Steve: She's in...

Peter: She was at Ohio Queen Breeders...

Steve: I think she's near Seattle now.

Peter: I know she's at another university. Her bee was the New World Carniolan, that was her big invention, but I've had lots of help from Sue Coby. A little bit of advice from Redmond Williams in Ireland, and basically, lots of people that I pestered many years ago that were only ever helpful, really helpful, especially Sue Coby and Joe Latshaw. My questions were never ending.

Steve: I can imagine.

Peter: It's like most of them said, if you can ride a bicycle you can ride a bicycle, but I can't show you how to do it. The advice was to get 50 virgin queens and practise, using condensed milk. You're going to kill most of them practising, and I practised. I spent a few days up with Michael Collier at Cornbrook Bee Farm, we spent four days up there.

For much of the season I haven't had time to be doing it but now with the boys doing more I'm starting to find more time, and last year more than ever. This season I've planned even more inseminations.

Steve: Did you say you need 8 micro-litres?

Peter: Yes, 8 to 10 micro-litres of semen per queen. First of all, you inject a tiny little bit of saline in, to show your tip is inserted correctly, because the tip is only going 1mm into her. You put the tip in about half a millimetre, then you have to bi-pass the valve fold, then another half a millimetre past that into the median oviduct, then you turn the syringe – the knob on the syringe – to drive in the saline. That shows you that you've got the needle in the right place and you then pump drone semen into the queen. You then back off slightly with the pressure as you withdraw the needle, and take the sting hook and the ventral hook out of the way, then that's it done. I take the queen out of the holder by blowing her out into my hand, mark her, clip her, put her in a cage, and about 15 minutes later she starts to come around – well, she starts to come around after 5 minutes, but it takes 15 minutes before she's mobile again, because she's just been zonked.

Steve: Then you have to knock her out again, don't you?

Peter: Yeah, you do that 24 hours before insemination or 24 hours afterwards.

Steve: Does it make a difference?

Peter: Yes, it does. It takes a few days for the drone semen to migrate to the spermatheca of the queen, and temperature is important in this as well as being active, hence putting a bit of brood in with the inseminated queen, because you know they're keeping that at 35°C. You want her to be warm and active, not cold. So, it's better...I've done it both ways and haven't seen a lot of difference, but it's supposed to be better to anaesthetise the queen for five to six minutes twenty-four hours before the insemination, because if you are doing the insemination and then going back the next day to collect the queen and anaesthetise her, that's all interrupting the migration of the semen into her spermatheca. You can end up with a queen that's perhaps not taken as much semen into her spermatheca as one that was done before. That's Sue Coby's reasoning on it, and I've got no reason to think that the best person in the world at this is wrong!

Steve: Seems fair.

Peter: But I've done it both ways and I haven't seen any difference.

Steve: When you've got an inseminated queen, that's basically going to be a breeder queen, is that right?

Peter: Yes, that's the general idea, but you won't know she's a breeder until you've tested her daughters. You raise daughters from her and open mate them (you can inseminate some as well if you like), but you open mate them, and then it's how those daughter queens perform that lets you know whether it's worth going back to raise more from the mother.

Steve: Yes, I see

Peter: It's all about daughter testing.

Steve: Does the inseminated queen last as long as one mated naturally?

Peter: Yes, sometimes longer.

Steve: I thought it would be less

Peter: As long as the procedure is done correctly they can be in full sized honey production colonies, no different to any naturally mated queen, and quite often they are better, for the simple reason that you know they've had a full dose of drone semen whereas a naturally mated queen may not have. If the weather is bad she may only get mated by two drones and a few months later she's a drone layer. The inseminated queen is not going to suffer that problem if she's properly inseminated. The only thing with an inseminated queen is that you need to keep an eye on them. They are best left in a small nuc for at least three weeks, then you can put them into a big colony. Because they are not giving off the full pheromone, and this applies even to a naturally mated queen, by putting them into a big colony first they may make supercedure cells. You just need to go through the colony once a week and break down any cells, and after about three weeks that stops.

Also, prior to inseminating the queen, the nuc she's in is queen excluded across the entrance, from the inside. The reason is that if you put it on the outside the bees will try to drag the virgin out to make her go on a mating flight. If you have the queen excluder on the outside they can get her cornered and they can damage her, but if you put a bit of slotted zinc excluder on the inside they can't get her cornered. Obviously, we want to stop her flying off to get mated naturally. After she's inseminated, until she comes into lay, which I find is usually between 4–7 days later, we want her to stay inside because she may not know she's inseminated and in nice hot weather she's going to attempt a mating flight. Because she's clipped she'll just get lost. The chances are she wouldn't attempt a mating flight but she might, so you leave the queen excluder on the entrance until she's actually laying eggs.

Steve: I guess until you've tested her daughters you keep her in a small nuc, because she may be very precious?

Peter: Once she's been in the mating nuc for about 3 weeks the next move for her is either a full-sized colony or generally into a six frame nuc. Then you can test her worker daughters, but more importantly you graft queens from her and test those. You are interested in her workers as well, but really the main thing is the daughter queens you produce from her and their worker daughters. Then it's repeatable. You go back to the inseminated queen and graft more. It's also easy to go back to the one I produced the inseminated queen from, and that drone line, so I can produce more of her.

Organisation and operations

Steve: It must be quite mind-blowing keeping track of all of this.

Peter: To me it's not; it's simple.

Steve: You've got all these different drone lines and different queens…

Peter: Most of that's in my head. I suppose it comes from my Grandfather. He was a shepherd and he had about 2,000 sheep. It's in my head, plus I'm a bit of a Mike Palmer; there's lots of notes on the hive roofs.

Steve: Right, on duct tape!

Peter: Duct tape and we write on the crown boards. The crown boards are like writing desks. There's writing over writing. I know what it means. The boys are often amazed; I walk into an apiary and I know the queens, I know their mothers and their grandmothers…

Steve: But you do write it down somewhere?!

Peter: I make rough notes and it's on the crown board, but a lot of it is in my head. I remember. I do have records as well, yes, for some, but the majority is in my head, Steve. I haven't got anything else to think about. [laughs]

Bees are all I think about, so I know who's who, and where they are from. I'm like that.

Steve: So, one day, probably many years from now, but one day when you've had enough, would you want one or more of your sons to pick it up?

Peter: Three of them are going to. Andrew's already been grafting and he's right into the queen production side of things. He produces queens to head the nucs that he's going to sell. That's what Andrew's interest is. Lewis is nineteen and that will be the next stage for him; he's had a go at everything.

Steve: The thing is, if you think about all the work, you're making these little silver things and…

Peter: That's all part of the enjoyment to me

Steve: But the amount of skill and the amount of time…and then all these different lines and all this testing, it's incredible. There are people selling queens who probably don't do anything like that.

Peter: I don't know.

Steve: I can't imagine many people are doing what you are.

Peter: A lot of people who are producing queens to sell just buy a breeder queen, Steve. They buy a breeder queen and then they graft daughters from her and open mate them. That's what the majority are doing. They say they are breeding queens; they are not breeding queens, they are producing queens.

Steve: Yes

Peter: There's a total difference between being a queen producer and a queen breeder. Breeding is when you are in control of the mating lines – total control – you know exactly what drones are being used to make which queens, that's total control of it. Other than that, you are just producing queens. Just producing queens is simple, simple as pie, but breeding them is a totally different thing. It involves a lot of time and a lot of observations, in other words, intuition. You need to observe how the bees are performing in all weathers; you want to know what they are like in good weather, bad weather, rainy weather, sunny weather – are they good foragers? What's their temperament like in all weathers? If it's a bit thundery do they come out and sting the hell out of you? Or are they still just calm and placid. This is all part of it. You can't get this just from sitting on a computer and imagining it. You've actually physically got to be out there going through your bees every day of the week, you've got to get to know your bees.

Sandra: He knows his bees like a farmer knows his sheep

Peter: I can't even rely on the boys coming back telling me what this colony or that one is like; I've got to see it for myself. I want to look at the queen and look at her brood, I want to know her temperament in all weather, how productive the bees are, how swarmy or not swarmy, and it's my intuition. It's basically one big guess then, but my intuition tells me that if I cross that queen with those drones I think I'm going to get a damn good bee. So, I do it. Nine times out of ten it works well; sometimes it doesn't. Nothing's guaranteed.

Steve: What happens when it doesn't work out well?

Peter: Well, you get bees that are absolutely awful.

Steve: Right

Peter: They might be too inbred, they might be unproductive, they might be stingy, but you soon delete it then; you don't repeat it.

Steve: No

Peter: Even Brother Adam made lots of mistakes over the years. It wasn't all perfect.

Steve: I'm sure, yeah

Peter: There were lots and lots of lines he had to delete because they weren't any good. Sometimes you can take two lines that are unlikely bees, one seeming gentle, placid, but not great honey producers and you can go to another line that is a bit more fiery, fantastic honey producers; well that could make a good combination. You are trying to combine the good points of both of them to make a good bee. So, you can have two lots of bees that are not necessarily very good. From observation you can think, "they are a load of rubbish" and this other lot are a load of rubbish too, but they each have qualities which may work well combined. That's done by intuition. Some other person would say, "why ever would I want to breed queens from that?"

How did you start?

Steve: For how many years have you been doing that?

Peter: We've been doing it for a living for over twenty years.

Steve: Are the first five years the hardest?

Peter: I don't know. I just did it as a hobby before, but I started off when I was very young. I had two great uncles that were bee keepers. One of them only had half a dozen hives in his garden, the other one had a farm and he had about forty hives. When I was seven years old I used to have to go along with my Mum every two weeks or so, and he was doing his bees. He never wore a bee suit either.

Steve: You were seven?

Peter: I was seven. He used to wear his trilby hat and shirt sleeves to go through the bees. I used to go along and that was my first experience of bees. That absolutely fascinated me then, but everything fascinated me then, Steve; every creepy crawly under the sun. I used to keep lizards, toads, frogs, newts, slow worms…I even came back with an adder in my pocket one day

Steve: Did you know it was an adder?

Peter: My father chopped it's head off. We had rabbits, chickens, ducks, geese, terrapins, fish – it was like a menagerie – we had every animal under the sun. Have you ever heard of a guy called Gerry Durrell? Gerald Durrell?

Steve: Yes, I think so

Peter: Well I was a bit like him when he was on Corfu at that age. "A Zoo in my Luggage" and "My family and other Animals" are some of the books he wrote, and he opened a zoo in Jersey. I was very much like that. I'd have trout in aquariums, I'd be out catching white slugs to feed to my toads, I kept everything.

Steve: I've found, with the people I have interviewed, that these people are all fascinated with nature.

Peter: I like all of nature. When I was a kid I was gone from first light in the morning, 'till dark. I used to like reading when I got back in the evenings; I've always liked reading. But I've always liked nature and being outside in it. I'd be in a river catching trout or picking up stones looking at the creepy crawlies under them, catching leeches, looking for newts, catching frogs, bringing back frog spawn, toad spawn, toads – everything to do with the wild. Climbing trees looking for pigeons' nests, buzzards' nests. That was my amusement as a kid. Now I see kids sat around getting totally bored out of their brains and if the computer goes offline they're dead!

Steve: Yes

Peter: That's just how it was in the countryside, anyway. I can't comment about cities or towns because I never lived in one.

Steve: No. Well, I grew up in the New Forest and we used to throw dried horse poo at each other.

Peter: We used to chuck sheep shit at each other [laughs].

Steve: It seems to be fairly standard.

Peter: Early on I went into the motor trade because I was interested in cars and motorbikes, but even then, I used to go off to the bee meetings, even when I didn't have bees. I still enjoyed going to the bee meetings, you know, the Association meetings.

Brother Adam's mating station and black bees

Steve: It's quite a small world isn't it? I mean you probably know a lot of the main beekeepers in the country.

Peter: Yeah, I know quite a lot. I know one out of this country, that's Mike; he spent a few days here. We picked him up from Heathrow and brought him back here. I think I took him to Dartmoor.

Steve: He showed me a photo of that packhorse bridge just down the road.

Peter: Yes. Then we sat here until 2am talking about bees.

Steve: He seems to be fascinated by a lot of what Brother Adam did

Peter: Yes

Steve: You've got that connection haven't you with Brother Adam? What's your relationship with the mating station?

Peter: I run it.

Steve: You own it?

Peter: I don't own it, no. My friend owns it, Anton Coaker who has something in common with us because he's a farmer but he does saw milling as well. I've known him for years. The mating station is 100yards from his house and his family have been there for five generations. The mating station was set up originally by Brother Adam through Anton's father, who leased it to Brother Adam. Well, it had lied derelict for years so I phoned up Anton and chatted to him about it and that's how I can now use it. We went down there and burnt loads of the hives and sterilised loads and stacked them in the shed. I took four boys down there with me and we spent the whole day there, that was back in 2012, and got everything sterilised. We were never going to use the old equipment, we use our own modern-day stuff. He'd had AFB there many decades ago and all the combs were there with pollen on, and the dead bees were there – they were just left there to die. There were just piles of dead bees; they left them there to starve to death. So, we wanted to get it cleaned up before we started using it.

I may, this season, have something to do with breeding black bees. Last year we made lots of bee hives for the B4Project.

Steve: That's for the black bee? Is it the Tamar Valley?

Peter: Yeah, well Mount Edgecumbe is where the bee-hives are at. It's a black bee sanctuary on the Rame Peninsula. Dennis and Sandra delivered the hives for them. Andrew Brown is their finance raising guy, and over a year ago he was talking to me about doing some queen insemination for them. So, I said yes. A couple of months before Christmas we were talking about using the mating station as well, so we'd be doing insemination and natural mating. I've also been talking to Dave Ledger and Phil Chandler. They all want somebody who can produce the Cornish black bee primarily to sell them to beekeepers in Cornwall, initially, but to produce them in commercial quantities; in other words, hundreds of queens.

Steve: Actually, it's quite hard to get hold of them right now isn't it?

Peter: This is why they want them to be more available, primarily to people in Cornwall but then to a wider area. They've got students at the University doing DNA testing on them, so they will be the right bees to breed from. They want them in reasonable numbers so they can sell them, so they will make money off them although that's not the main object, because they raise money through grants and other funding, which is what Andrew Brown does. He seems to raise money from here there and everywhere for the B4project.

Steve: The main idea as I understand it is to flood the area isn't it, so that they become the local bee.

Peter: Yes, this is why they want them to be supplied to people in Cornwall to start with.

Steve: What about acarine, aren't black bees susceptible to that?

Peter: I haven't seen acarine for years

Steve: Maybe it's lurking?

Peter: No. I don't think so. I think ever since we've had varroa...

Steve: So the varroa treatment knocks it out?

Peter: That's right, yeah. I haven't seen Braula either; we used to see those frequently, sat on the back of bees. Sometimes they'd be crawling all over the queen and you used to have to blow smoke on them to get them off. They weren't harming her, just cluttering her up. They were there because they were being fed. The queen's being fed so they home in on the queen. They were quite harmless, apart from if you were doing comb honey because they would make little trails under the wax cappings, and spoil the appearance. I haven't seen those since we've had varroa, around 1992–93.

Steve: And the same for acarine?

Peter: The acarine...we used to have a craze for the bright yellow New Zealand bees, Exeter Bees used to sell them, everybody was buying them in. They are a lovely gentle bee, like little moths flying around, gentle, bright yellow, and what happened in this country was that they'd never stop laying. They'd build up massive colonies, collect a load of honey, then they eat it all and die of acarine before the winter! That was happening to everybody that got them, they were spread around all over the grass, they just die of acarine.

Steve: When was this?

Peter: Oh, back in the nineties. 1993–94 I suppose.

Steve: I've got a picture of one of them on my phone from my recent trip to New Zealand...is that them?

Peter: Yeah, that's exactly them, Steve. New Zealand Italians. They were more yellow than the Italian Italians. Ever such a gentle little bee, you could do anything with them. They used to build up massive colonies but, in our weather, they would eat the honey, and one day you'd find them all spread out over the grass just crawling around with acarine. Clumps of them on blades of grass...it was happening to them all over the country. We only went on with this for about 3 years then we stopped buying them. They were very susceptible to acarine; the other bees weren't getting it at all, just the New Zealand Italians. So that was the end of that; no more of those.

What Murray used to buy was the New Zealand Carniolan. There was a big story about how they did that; they imported Carniolan drone semen and inseminated their queens and eventually turned their Italians into Carniolans, by raising daughters and inseminating them and so on. Over several generations they turned their Italian bee into a Carniolan. You can read all about it. That's what Murray [McGregor] was importing in packages at one time. I have no experience of those myself because I've never had them.

Steve: Both him and Jolanta both say Carniolans are the best for where they are. He doesn't care what type of bee it is.

Peter: So long as they are collecting honey!

Steve: Yes, but it so happens that the Carniolan type of bee does best in his conditions. His whole operation is geared up towards the heather.

Peter: Well most of ours here are mainly for the heather, but we do have good flows before that, whereas Murray is totally for the heather really isn't he?

Steve: Yeah. But the black bee thing; are you quite excited by that?

Peter: I don't know if I'm excited. I've got some black bees. I got some from three breeders in Ireland, from Mac Giolla Coda, the Galtee black ones, some from Gerard Coyne in Connemara, and some from John Summers; they are all black bee fanatics in Ireland.

I also have two or three French black Amm bees. One of them was evil and one of them, which I've still got, is over at Ebbs Hill. The intention is to graft daughters from her, they are quite violent, and mate her daughters to the Galtee drones, because they are quite gentle. It's going to take a few years of this black bee breeding but eventually I want to incorporate it into some of the Buckfast lines. Brother Adam used French black bees for several years, about ten to thirteen years. Looking back on the pedigrees he was using Alec Gale's French Amm breeders in his Buckfast lines, it's all on the internet on the pedigree. Alec Gale, as in Gales Honey, used the French black bees for his honey production. Brother Adam was best buddies with guys like Alec Gayle and Manley back in those days, and he got his French black bees through Alec Gale, to cross to his Buckfast lines at Buckfast Abbey. I want to put a bit of that black bee back into them again.

Steve: Is your aim to calm it down, temperament wise?

Peter: Calm it down temperament wise, and put some of the black bee hardiness back into the Buckfast line, that's my aim.

Steve: Sounds great.

Peter: I'm interested in the Cornish black bees as well, and may incorporate some of those. I'm just going to be doing it, observing, and seeing what I come up with.

Steve: You use the isolated station on Dartmoor but I think you've got another one on Exmoor haven't you?

Peter: Yes, that one's much closer obviously.

Original Brother Adam mating hive

Steve: Is that jutting out to sea or something?

Peter: No, it's right in the middle of the moors, a Godforsaken place where you wouldn't think a butterfly could live. When you take the bees there you have to feed them constantly because there's nothing there for them, hence no other bees are going to go there.

Steve: Makes sense

Peter: ...Until heather time. I have to get the mating done there before the heather flowers in August. Before then it's just a desolate wasteland, but once the heather comes other people might move their bees onto the moorland, like Quince Honey Farm and people as far away as London – they bring bee hives down here on Exmoor for the heather. We have thousands of miles of heather. They come from Gloucestershire, Salisbury, London, all over the place.

Steve: Do they have to get permission?

Peter: Yes, they get permission. I've got established sites, obviously; I was born here, although I lived in Scotland for a while many years ago, and when I was eleven I went to school in Chipping Norton – have you heard of Chipping Norton?

Peter in the workshop

Steve: Yes, I've heard of it…

Peter: My mother is still there. She's 85 or 86 now or whatever; she's quite fit and has two horses and walks about five miles a day, and mucks her horses out.

Steve: Fantastic

Peter: She bought a place there about 45 years ago and she's been there ever since.

Steve: I've just been reading a book called Border Bees by Colin Weightman, have you heard of that?

Peter: I don't think so.

Neonicotinoids and pesticides

Steve: What do you think about neonicotinoids?

Peter: What do I think? I hate pesticides with a vengeance. I can't imagine any beekeeper in their right mind liking pesticides. As regards the neonicotinoids that they have now, what worries me is what will take the place of it? I haven't seen any harm to my bees, but I live in a place that's not arable. This is wild countryside, we haven't really got anything arable. All of my bees are on wildflowers and chestnut trees and rosebay and clover; it's not like some people who have nothing but thousands of acres of oilseed rape. Even when I've taken bees to oilseed rape over the years I've never had any problems, apart from 32 hives which were poisoned, but that was a foliar spray, nothing to do with neonicotinoids. In fact, I wish they'd been using neonics back then. Apart from that one incident, all of the times I've had my bees working on oilseed rape I've never seen any ill effects in any way whatsoever; no harm to the bees, no reduction in the colonies, no after effects, no ill effects, nothing wrong whatsoever.

What really worries me if that's all banned, which it is, is that they start going back to the old foliar spray. They don't always stick to doing it at night time, because when they've got thousands of acres to do they can't keep coming back every night to do it. When they get spray contractors in the contractors are going to spray that crop. It's like they did with the lambda-cyhalothrin and triazole; one thing on its own is safe enough, but when they tank mix it the triazole forms the lamba-cyhalothryn into an emulsion, and according to Syngenta you need to keep your bees shut up for three days, at least. It's no good overnight, you need to either move them away or shut them up for three days, because that emulsion that forms on the plant is toxic. The fungicide mixed with the insecticide together is the problem; if they used one on its own it would be safe the next morning. They mix the two together because it saves time and money – why do two separate sprays when you can chuck it all in one tank and do it at one time? It's the economics of it.

Being a beekeeper, I hate pesticides but I really worry; better the devil you know...if we go back to some of those old pesticides that were used, and I've seen the result of that – piles of dead bees outside hives – that's what worries me, it's really indiscriminate stuff.

Steve: Yeah.

Peter: But on the whole I hate all of them

Steve: Fair enough.

Peter: But I live in the real world and I know that agriculture is going to use them. They are not just going to let all their crops be eaten by insects, no more than I'm going to leave my bees to be eaten by varroa mites.

Steve: That law banning neonics, it's across Europe isn't it? Do you know...I mean, somebody must have said these things are killing bees, or pollinators, for the law to have been passed.

Peter: Well, they're killing wild pollinators, yes. Bumble bees and butterflies and so on.

Steve: So it's more about protecting other pollinators than honeybees.

Peter: I would say so, yes. Because I haven't seen any ill effects on honeybees. All I see is the numbers of honeybees going up and up and up. No doubt the research says it's doing harm to wild pollinators, yes.

Steve: You wonder about the people who make policy – who do they listen to? Presumably they would listen to beekeeping associations and farmers but I don't know.

Peter: I'm not very fond of them, but then again I'm not very fond of diesel particulates either, and I'm not very fond of aeroplanes spewing out exhaust all over the country. I think all aeroplanes should be grounded and turned into saucepans. The cities are all getting polluted with diesel particulates and petrol, the aeroplanes are up there spewing thousands of tons at a time down onto us. I know they did some research years ago about how clean the air was when there was an aeroplane ban for a few days when that volcano was erupting. It was amazing how clean the air became just from a few days of no air traffic.

Mentors

Steve: So back to bees, you said you had an uncle that kept bees?

Peter: Two great uncles.

Steve: Were they your mentors at the beginning?

Peter: Yeah, one was anyway, the other one…I used to go and stir his bees up with a stick. I wasn't too fond of him.

Steve: I bet he loved you!

Peter: I used to make a point of it every time I passed them, which was about three times a day. Anyway, do you want to see where we make our beehives?

Elaine: Wait a minute – I'm interested in who mentored you

Peter: It was mostly me, I loved wildlife in general and bees in particular. My great uncle helped a bit but it was mostly me.

Sandra: He started off with one beehive at the bottom of the garden. He used to stare at it for days and days.

Peter: I had them about two fields away then, didn't I? One day I spent six hours just watching bees at the entrance. I don't mean sat in a chair, I mean I was lying down with my face at the hive entrance, studying bees. I got ran over by a couple of hundred shrews that day. I saw a shrew explosion, which I'd

Hives in winter

never seen before nor since. I had heard they happen. Literally hundreds of these shrews, apparently they are like lemmings, they have a population explosion every so often, and they go on a migration. When I was lying there studying bees for six hours, it was a nice day obviously, I wasn't lying there in the rain, and these shrews were going over my legs, my feet, and everywhere. There were hundreds and hundreds of these shrews going through this woodland. This went on for, like ten minutes, just shrews going through, going through, going through and then it all went quiet and they were gone. Literally countless hundreds of the things. I've never seen so many shrews. It was like one of these lemming or wildebeest migrations but they were shrews!

Steve: So you stopped looking at the bees for a while at that point?

Peter: Yeah, I was watching the shrews then, Steve [laughs]

Queen grafting

Steve: When I spoke to Jolanta she said that it has to be the Chinese grafting tool, the one with the bamboo reed not plastic. Everybody has their favourites, don't they?

Peter: Everybody has their favourite, I've used a cocktail stick, a blade of grass, I've used most things, but for the last twenty years I've used a 000-sable paintbrush, what I call "three hairs and some air." I like using new comb for grafting, not old black comb; I think with old black comb it's better if you are using a stainless-steel tool, but for a brush I like comb to be as new as possible. To me the eggs show up better and also the comb is nice and soft, and I do all of my grafting sat in the front of my truck.

Steve: Do you have some headgear on or a torch or something.

Peter: Just a light, it costs about £80, it's quite expensive but it's a quartz light. I had a different pair of glasses made for me at the opticians just for grafting. They are a bit stronger than normal. I just wear the head light, glasses and hold the little sable brush. I hold the grafting bar and the frame in the same hand. I have the grafting bar which has ten cells on it, which I sort of hold with my thumb, and the frame at the same time. The cell bar rests on the top of the frame in my hand. I just sit in the truck. I use one end of the brush to flatten the cell walls, flip the brush around and then I just lift the larvae off and into a cup, and keep going until the bar is filled. That bar goes down on the seat with a damp cloth over it, I pick up the next bar, do ten more, then they are both clipped into a frame and put in the cell builder. That's how I do all of the grafting. It takes about ten minutes to do twenty cells.

I think Mike does most of his grafting in his truck.

Steve: Yes, and he listens to music whilst doing it.

Peter: I have no music but I do it all sat in the front of the truck just resting on the steering wheel.

Steve: He uses queenless starter colonies.

Peter: I do a slight variation on that; I don't bother to remove the queen but I do separate her, and I do divert all of the foragers to the top section. I have something like a Demaree, with the queen in the bottom box but I open up an entrance at the back of the floor and close the front entrance. I have another floor above that box, a split board, with the entrance facing the front, and double brood boxes on that, very strong, loads of brood, then a queen excluder and supers. All of the foragers in the bottom box will fly out of the back and when they come back they will go to the front entrance and end up in the top. That's the cell builder – they build the cells, but I do a slight variation from Mike, because after five days all of our cells come out and go into an incubator, and that frees the cell builder up to do more.

Like Mike I have lots of colonies doing this so we can get the incubator filled up with lots of cells. Then we block the back entrance and open up the front entrance on the bottom box. We leave the top entrance still, because a lot of bees are still using that, and take out the split board in the middle, so they are a queen right colony again. It's the same idea except that Mike takes the bottom box away. I leave it there and just alter the entrances.

Steve: Sure.

Peter: The top entrance is always open to allow any drones to fly out. All I'm doing is opening the back entrance, blocking the front and blocking off any access through the middle, so the foragers coming back with pollen are going into the cell builder above, which has no queen, no larvae; just sealed brood and emerging baby bees, plenty of pollen and the gap for the cell bar.

Steve: Is the main reason for diverting the foragers to get the pollen?

Peter: Yes. They are bringing back nectar and pollen which is feeding the bees. Sometimes if I get a colony that's in swarming mode, I take the queen away and break down all the queen cells. I keep the royal jelly from those because I always wet graft, I don't dry graft, I like wet grafting...

Steve: Is that water plus jelly?

Peter: No, I just use royal jelly, I don't mix water with it. I keep it in little glass tubes and freeze it. When I want to graft I put the tube in my shirt pocket so that it warms up and then I put a tiny dob in each cell. I've done dry grafting but I just prefer wet grafting. I put a little jelly in each cell then a larva on top of it.

But with the swarming colony, what they want to do is make queen cells. They are in prime condition for doing it and are strong enough which is why they're swarming, so I take the queen away and break down all queen cells, wait until they have no open brood, and give them twenty grafts to work on.

Steve: I bet they love it don't they?

Peter: Yeah, they love it. They will only ever do one, maybe two sets, and if you try to do more the quality goes downhill rapidly; I wouldn't do it. I prefer to just let them do twenty grafts and I leave them right up until a couple of days before they are due to emerge, then bring them back to the incubator or put them into mating nucs. During that period they get past their swarming fever. They are broodless, because it will all have emerged, so you can give them the mated queen back again.

All about swarm control

Steve: What about swarm control?

Peter: It all works to the nectar flows as to what you do at what time of the year. We take some nuc boxes out to the apiary and when we find a big colony in swarming mode we just pull out a couple of frames with the queen, put them in a nuc with a shake of bees and some stores, then go through the main colony shaking bees off every frame and breaking down every queen cell. We notice whether or not they are in swarming mode by the size of the queen. When she has slimmed right down and is starting to run around and her egg laying has reduced they are going to swarm; if she's still big and fat and laying like mad you can usually get away with breaking the cells down and giving some more space. We leave it until the next week – we inspect every week. The nuc with the queen goes to another apiary. If she's still good she'll come back but if not she can build up the nuc then she'll be squashed and a young queen put in.

Steve: You've just taken out the queen and knocked down all the cells, so there's nothing?

Peter: Nothing except eggs and larvae, and bees, yes.

Steve: So they make emergency cells.

Peter: Yep. They can't swarm because they haven't got a queen. All they can do is make more queen cells. We go back the following week, shake all the bees off the frames, break down every emergency cell. We then take a frame with larvae and eggs on it from another colony, mark the frame with an "X", and put it into the hive and leave it for another week. After that we either take a young mated clipped queen back with us – a new queen – or we go to the nuc with the old queen, and now instead of being skinny she'll have fattened back up again and will be back to laying, and if she's a good queen we re-cage her and go back to the colony that was swarming.

Apart from that one frame we gave them they have no brood, so you've now only got to inspect that one frame. Their swarming impulse is passed, long passed, so you take that one frame out, shake the bees and destroy all cells, then put the queen back in there, whether it's the new queen or the old one. Generally, that's it for the rest of the season then. They settle down, she's now got the whole box to lay back up again, plenty of space, the number of bees has dropped as the older ones have died off, and the swarming impulse is gone.

Steve: That really interests me because the splitting thing, the artificial swarm methods, I think maybe I'm not very good at it but I always end up with lots of small colonies and no honey.

Peter: Yes, this keeps it as a big powerful colony. If you want to boost it back up again afterwards you've got other colonies so you can just plonk a couple more frames of brood in there. They usually settle down and you have no more swarming problem from that hive for the rest of the season.

It's a bit like the German method, a reverse artificial swarm. You put all empty combs in the bottom box, with one frame of brood, which you mark. You artificially swarm them, vertically, but instead of the traditional way when you put the queen in the bottom box to build up the nest, this way you don't. You leave all the foragers in the bottom box with just that one frame of brood and no queen. The queen goes in the top box. The older bees leave the top box and fly to the bottom, so they aren't going to swarm from the top because they've got no flying bees; it's just baby bees up there with the queen. Any queen cells up there, they'll generally destroy them themselves, but you can do it if you like. The queen will start laying up there again and you do this over two cycles. After a week you pull the marked frame out of the bottom box, shake the bees off and remove any queen cells, then give them another frame of brood from the top box. The forager bees are still working like mad collecting honey. They are queenless and trying to make a queen from the one frame of brood...after the second week you break down the cells on the one frame and unite the boxes together. You are back to a colony that has got past its swarming impulse.

We are doing the same sort of thing except we take the queen away and we get a new nuc out of it.

Stopping the wax moth

Steve: What about storing drawn comb? You are adding boxes and getting foundation drawn in the summer but in the winter when you take it away what do you do with it?

Peter: We stack them up in a shed and fumigate them with sulphur, then we wrap them up in polythene and that's it. We don't check them again until spring and we've never had a problem. We have lots of greater wax moth around here as well, but that sulphur treatment does the trick. It's much better than using acetic acid, which burns everything that's metal.

Steve: Where do you get that from then, the sulphur?

Peter: From the garden centre.

Steve: Like a greenhouse fumigator.

Peter: Yes. You can get ripped off by a bee equipment retailer or just get it for about a quarter of the price from the garden centre.

Steve: That's been one of my problems; wax moths.

Peter: We have loads of them. Sandra goes round whacking them with a bat in the autumn because they are everywhere. There's never a sign of them in the combs in the spring. I used to use old deep freezers years ago. You need to either freeze or fumigate the boxes as well as the combs because the eggs can be in the boxes too. The sulphur wipes the lot of them out, the big ones, the small ones and the eggs. We stack them up and fumigate about 200 at a time.

Good years, bad years

Steve: Great. How about amazing and terrible years?

Peter: I think one of the hardest winters was probably 2010 but one of the worst years I've ever known was 2012 and the knock on effect into 2013. Of recent years that was the worst. It rained from 1st April until the third week in July in 2012, followed by the coldest late spring for 50 years in 2013. That's when we lost hundreds of colonies in this country. Some bee farmers lost 800–1,000 colonies.

Steve: Is that the one where it got bad in Scotland?

Peter: Yes, they lost hundreds. There was one bee farmer there who had 10 hives left out of 800. A local honey farm had lost 700 hives by March.

Steve: Was it starvation or something?

Peter: No, it was the lack of pollen basically, because in 2012 it rained constantly. We had bees out on the hills collecting nothing. Luckily down here in the valley is a micro climate and bees were filling supers. We loaded those bees up from several apiaries over the hills and very rapidly brought them down here and we were just plonking them in fields all over the place. There were hives all the way down the hedges and everywhere.

It was like a lottery though, because in some parts of the country you couldn't do that, we were lucky. In a few weeks just about every colony that had been close to starvation was filling two to three supers, so it was a brilliant recovery which carried on then onto the heather afterwards. In a lot of places, it just kept raining and bees couldn't collect enough pollen, so they weren't raising brood, then it got to the autumn when they are supposed to raise the winter bees. They couldn't collect enough pollen to raise enough winter bees to get through winter. The colonies went into winter with predominantly older bees. In January and February it was quite mild but because the old bees had dwindled, colonies were very tiny in March, maybe just a couple of cups of bees in a lot of parts of the country, and we had that here in the ones we left up on the hills.

Then at the end of March along came the coldest late spring for 50 years – it froze almost until June. Well, that was the death knell, the final straw for all of these little colonies that had made it that far. They still couldn't get pollen because of the cold, they couldn't raise young bees, so they died, by the thousands.

Steve: That wasn't a pesticide or anything related to human activity.

Peter: No, purely Nature, weather. It was one bad year with a knock-on effect of a cold late spring the following year. If it had been a warm spring even the tiny colonies may have made it, but that was just the final nail in the coffin.

Steve: It didn't hit you personally as hard as some?

Peter: No because luckily, we had the micro climate in this valley. We had some respite in the third week of July but up until then up on the hills we were just feeding bees all the way through summer. In the springtime we had colonies with only a frame or two of bees, so we found the colonies that were booming and shook bees off into Correx boxes, like packages, and used them to boost the weaker ones. That's how we saved a lot of our bees. Luckily, we had plenty of hives and plenty of strong ones to boost the weak ones.

In Scotland they were losing them by the thousand. That's when Murray, I believe, first started importing packages. The Scottish government gave their bee farmers money for re-stocking. I believe that Murray, among others, was central to doing that. He wasn't just doing it to help himself; he was helping other beekeepers and there are beekeepers all over the country who are grateful to Murray for that, because they would have been out of business. That's how it was.

Steve: What about amazingly good years that you can think of?

Peter: In recent years, 2014.

Steve: You had a terrible year, then a terrible year, then a great year?

Peter: A great year, yes. It couldn't have been any better. As fast as you could put supers on the hives, no matter where they were, they were filling them. It was good right the way through from the spring up to the heather.

The worst heather year was in 2015. We had bees on the heather and it flowered all the way through August, there was a chilly wind and some rain, but for some reason the bees got nothing. This went on right through September, and they started collecting honey in October. Then they had a mega honey flow for ten days, they all filled up one or two supers, and that was the end of it. That was the latest season ever. Murray had the same. He took his bees to the heather in July and they sat there right into September, and the colonies were dwindling and dwindling and dwindling, collecting nothing and needing to be fed, and he said at the time that between him and his father they had never known that happen in 60 years. I've never known it. It was in flower all the way through August but the bees collected nothing, then all of a sudden right at the end there was this mega honey flow. I'd never even known bees to get much off the heather in September, it still flowers but they don't usually collect anything after August. It was a freak of nature.

Steve: You hung in there though, you didn't bring them back.

Peter: There was nothing to bring them back for, except feeding them, Steve. They were there for the heather. It certainly wasn't a good season though, they just filled up one or two supers. There was none of this like last season and the year before. 2016 was a super-duper heather year, as was this last one in 2017. 2016 was probably a bit better but it was early. I think we had nine or ten tons of the stuff, just heather, and I should think Murray would have had about 60 tons with all his thousands.

Honey production and coping with robbing

Steve: When you finally do get full supers do you leave it on the hives and bring it down all at once?

Peter: Because a lot of our apiaries are fixed we de-super them up on site. Generally, we go round and clear all the bees on site and bring just the supers back. If it's going for cut comb, we try to get those supers off just after the bees have sealed it. For the rest of it that's just being agitated and extracted it stays on the hives until the end of the honey flow. We generally clear them on site but in the past we have moved them back down to their winter apiaries with supers on, it's just heavy damn work.

Steve: That's the harder way to do it.

Peter: Yes. Normally we go out with the trucks and I take the blower with me. We put clearers on...

Steve: Do you use that spray that bees hate the smell of?

Peter: I have got that but I don't like it. Mike uses that I believe, it's like an almond smell. I have tried it a couple of times and it hasn't been very successful so we just use the vortex type rapid clearer boards with two holes for the bees to go down through and they clear a couple of supers in about three or four hours. For any stragglers we have the blower. We just blow them out by the hive entrance, stack them and sheet them over. One of the boys will be on the back of the truck, two boys will be carrying the supers and I'll be there with the blower. We stack them and sheet them over.

Nearby Bossington

Steve: That's to stop robbing?

Peter: Oh yes, you will get robbing, depending on the weather. It's not so bad at the heather sites because it's getting cooler, so you don't get too much robbing there. You get more robbing problems clearing the summer honey when it's come to the end of the flow. Also we get problems inspecting colonies that aren't on the heather in late August, or trying to make up nucs. If you are not really quick and have the hive open the bees are all homing in on it and trying to rob it out. Sometimes you have to work really fast or just forget it; leave it for a while. Sometimes we wait until it's almost dark because if you open them up in the daytime it's mayhem, they can go robbing crazy. Once that gets started you can't stop it. They'll be piling in the entrances of hives, so we try to avoid that at all costs. Sometimes if we are working in a apiary and they're getting lively we just shut up and bugger off to another apiary, then we'll go back in the evening.

Sometimes you get hives that are robbing out nucs. We'll either move the nucs to another apiary or move the robbing hive. If you go there late in the evening you will notice that it's usually only one hive that's doing the robbing. They are not fighting; they are going into the nuc and taking its food and the nuc bees are letting them do it. You put 3 litres of syrup into a nuc to feed it and two days later you lift the nuc and it's as light as a feather, you look in the combs and they are empty, right? So you give them another 3 litres of syrup and the same happens, so you study it, you watch it. You can see that all the other hives are quiet, and this is at 8.30pm, but there's one which is flying like mad, all the bees on the front of the nuc. You lift the hive that's doing the robbing – they weigh a ton, yet they've only just been de-supered a week ago. They had nothing and now they are suddenly full of food, so they are obviously the culprit. We just block the robber hive up, strap it up, put it on the back of the truck and move the robbers to another apiary.

If we suspect that it's more than one hive doing the robbing then we'll load up the nucs and we'll take them to another apiary and then give them a feed, because once it's started you'll never stop it. You've either got to move the robbing hive or the nucs.

Steve: That's interesting because it's happened to me and I blamed wasps, but it could have been robber bees.

Peter: If it's a nuc a good way to boost it up is to move them in the middle of the day when they are being robbed, and they are full of the robber bees. Block them up and then move them because the robbers are trapped inside, and when you move them to a location two or three miles away the robber bees automatically become part of that colony. They come out and think, "Oh, where am I?"

Steve: Yes, "we might as well join this lot."

Peter: They can't find their way back so they have to join the lot they were robbing. There's all these little tricks. I had two six frame nucs and one nuc was robbing the other, so I just moved one nuc away. It was 9pm and out of ten nucs there were two that were busy, one where they were all round the front and the other one you could see them flying back and forth, so I just picked up the one that was being robbed, moved it, and that was the end of the problem. That was last autumn.

Steve: It's massively helpful to have more than one apiary isn't it?

Peter: Yes.

Advice for those who want to become commercial

Steve: OK, I always ask this one and always get a similar answer but I'll ask it anyway…starting out as a commercial beekeeper…I mean, typically someone is a hobby beekeeper with ten hives, it's all going well, and they think they should quit their job and become a bee farmer, because it looks great…have you got any words of wisdom or advice for people who are in that position?

Peter: Yeah, do it slowly. That's my word of wisdom for that. Don't jump in at the deep end, make sure it's the life for you first, keep hold of your day job to start with, and build up your beekeeping business until you have enough hives to make a living from them. Never have all of your eggs in one basket. There's lots of wide-eyed people have gone into it, bought themselves 200 hives and they find out, "This is not easy, this is not the life for me." It's also very expensive to do it that way. Build up your business slowly and buy stuff as cheaply as possible because you can soon spend a fortune and be bankrupt.

I don't know what other people's advice is?

Steve: Pretty well that, yeah, that's about right.

Those who have inspired me

How about heroes? Have you got any heroes, any people who you thought, "Wow, I like what they are doing."?

Peter: Yes. I thought Brother Adam was rather wonderful and R.O.B Manley was one of my favourites, of old-time beekeepers; he's my hero I'd say, R.O.B Manley.

Steve: My favourite beekeeping book is Honey Farming by Manley

Peter: It's my favourite as well, Steve. But Brother Adam, obviously, as a bee breeder – he was a very clever man, but most of it was intuition. That's what beekeeping is; you've got to have good intuition, a good feeling and be studying and watching bees, you've got to be in touch with them. You can't be remote from them. You've got to be hands-on and you've got to get to know them. That's what Brother Adam was like and that's what I'm like too, in that respect.

So they are my heroes, and probably Sue Coby as well.

Steve: Have you ever met her?

Peter: No, I haven't. I've chatted with her, I've had lots of emails back and forth, but I admire what she does and her skills. But of all-time honey farmers it would definitely be R.O.B Manley. I like his way of writing, I like his attitude, he's very blunt, especially against certain people, but he writes like a real person. I can relate to him, he's real.

Buckfast Abbey and Brother Adam

Steve: What do you think about Buckfast Abbey and how it's changed? It's more educational now, isn''t it?

Peter: Yeah, they've just got a few hives down there, which are for training new beekeepers. That's what Claire does; she just trains new beekeepers.

Denis and I were down there to pick up Brother Adam's insemination apparatus, which is now in the Netherlands.

Steve: Oh right. Is that going in a museum or something?

Peter: No, no, the Marken mating station. It's a small island and it's a Buckfast mating station. They came over for a couple of days to meet me and I took them to the mating station on Dartmoor and they went back with two or three of Brother Adam's old hives, the mating hives, which they renovated. They mate queens from all over Europe on Marken, Buckfasts, and they built a few reproduction hives copying the old originals. They'd arranged to have Brother Adam's insemination apparatus so I went out to pick it up and I also picked up his automatic honey bottling machine which weighs a ton. It's made of cast iron and has all weights and electrical gizmos and relays on it – I've got that out there in the shed. It was made in 1922 I think, and it was first repaired in 1940, because I've got letters from Brother Adam to the place up in Wigan where they bought it. They managed to get the parts, in wartime, down to Buckfast Abbey to repair the machine within two days!

Steve: Pretty impressive.

Peter: Really impressive. That was 1940. From 1922 when they bought it to 1939 when it first went wrong...the letter from Brother Adam to the manufacturers says what the problem was, and I have another letter from the company saying that they had sent it back. It was two to three days. I don't know if they were telegrams or letters; I've got them anyway, and all the blueprints of the machine as well, plus the bottles of mercury that go in it.

Steve: Was that easy to get hold of?

Peter: What, the machine?

Steve: The letters.

Peter: Well, I have got them but I haven't a clue where I've put them. They are away in a box somewhere, Steve, but they are somewhere. He's called the Reverend Brother Adam in those letters.

Steve: You breed Buckfast bees too don't you?

Peter: Yes, I breed Buckfast queens but I have nothing to do with Buckfast Abbey.

Steve: No

Peter: Because the Abbey had the mating station; well, there were two mating stations, but the Abbey didn't like Brother Adam, you see?

Steve: He probably went a bit too heavy on the bees for them, maybe.

Peter: They lost an ideal opportunity there, they chucked it out the window. They chucked all that stuff in the pig shed – all of his trophies, stuff that he'd been awarded over decades, pewter plates, silver, cups, awards, his insemination apparatus, frames – they took it all and chucked it in a pig shed. Most of it went into a skip. It's almost like they disowned the guy. His best friend, Pete Donovan, who used to work with him, visited. I knew Peter as well.

They did nothing to help Brother Adam. He died in an old people's home just down the road. He was disowned by that Abbey, really, because he was famous. Of course, being a Benedictine monk, they don't really want you to be famous, but he made them a lot of money and he was famous. Maybe they resented that. They tore out all of his honey tanks. They are only interested in making Buckfast wine now. They could have a museum to the man. With all the stuff they had they could have opened up a museum to a world-famous beekeeper. What have they done? Chucked it away. Some of his papers are locked in a vault, never to be seen again.

Most of the European Buckfast breeders, the Germans, the Dutch, are absolutely disgusted with Buckfast Abbey. They have nothing good to say about them, because they are disgusted with the way they treated Brother Adam. I can't say any more about it.

Steve: It is a shame, isn't it?

Peter: They don't breed Buckfast bees there. They've got local bees and when I was there, they had half a dozen queens imported from Greece. It's not Claire's fault; it's the fault of the Abbey.

They had the mating station and they just left all the bees up there to starve to death. That's how much they thought of it. The Abbey basically walked away from the mating station, they wanted no more to do with it. There's a place there where he used to raise 420 queens a year, all the mating boxes, his shed… Dennis has got his ladder, Brother Adam's ladder; it was made in 1940. They used to use it to go up the apple trees in the orchard. They walked away from all of this, they just deserted it. When we went there, there were piles of dead bees in the hives.

Which bees are 'best'?

Steve: He wasn't universally loved, I guess, because there's this thing about different types of bee.

Peter: I think he had a good following with his type of bee. You've got zealots, I would call them, of all different bees. Some are black bee zealots, some are Buckfast zealots, but Brother Adam was not a zealot of any bee, for the obvious reason that he had all of them. He crossed all of them. It was a mixture of 13 different sub-species. So he loved all bees. Nowadays you've got Buckfast zealots, black bee zealots, Italian queen bee zealots, Carniolan zealots…and that's what they really are, and then you've got other people like myself, and others obviously, that actually like all bees. You'll come across these people that are like, "Ah, the only good bee in the world is a black bee." All imports are rubbish, nothing else is any good…they are nothing but zealots, really. If they were real lovers of honey bees they would like all of them.

Steve: OK

Peter: Brother Adam had black bees, obviously, because he crossed them with his Italians. To do that you've also got to keep them in their pure strain, because if you don't keep the pure strain of the sub-species you can't do further breeding. You've got to be as interested in the pure sub-species as in what you are crossing, they are all an important part of it. Without one, you can't do the other, Steve.

How can your average bee keeper improve his or her stock?

Steve: So with you, obviously you have all of these amazing queens, and you have carefully selected them and so forth. For some modest hobby beekeeper, they can always breed from their best queen but they've only got ten colonies or whatever, is it feasible for them to improve their stock?

Peter: Of course it is! I think all beekeepers should be raising their own queens. Just select the best, always select the best. Unless you've got somewhere isolated or an inseminator you haven't got the control over the mating, but by always selecting from the best you are making slow headway. You can make good headway over time and can produce good queens. You haven't got to keep buying queens. You can be self-sustainable with your own bees just by putting the time in and selecting, and hopefully spreading your bees around your local area. The more you can get together in your local area with like-minded people who want to produce as good bees as they can, the quicker it improves.

Steve: I guess if everyone did it you wouldn't be able to sell your queens.

Peter: I try to talk people out of buying queens quite often. I'm not a salesman. I like beekeepers to be self-sustainable, to manage with what they've got. That's what I've always done and it's why I do what I do. When we have these apiary meetings and all the local beekeepers turn up I show them how to do all of this because I'm trying to encourage them to do the same as me. I want them to go away and graft some cells and produce some queens.

When I'm doing these quad boxes, the over wintering of the queens, I do it on a large scale but somebody could do it on a small scale – they could just have one quad box and overwinter four queens. You can be a hobby beekeeper with a few hives and still do that. In the springtime you've got a backup there; you've got queens available early. You can use them for re-queening, replacing drone layers, you can make up nucs with them, let friends have them, you can sell them to people, but you haven't got to get on the telephone and order a queen or get one from Greece or Lord knows where. You can become independent.

How do the quad boxes work?

Steve: Those quad boxes have baby frames in them, don't they?

Peter: Mine are half length. They are National shallow frames, Hoffman spaced, and they are half the length of a normal frame, but the same depth. Each section of the quad box holds five frames. In the summer we split these quad boxes to make up individual mini nucs. The individual boxes hold three "half-frames". We go to the quad boxes, and if they are strong we'll split them into either two or three individual boxes depending on how much brood they have. As long as they've got three frames of brood we'll split them into two, so each three-frame box has one frame of brood plus two spare drawn combs added. We'll leave the queen in the quad box, or she may have been sold, but we'll still split them up.

We end up with a truck-load of individual three frame boxes with one kilo of fondant in. They get taken to another apiary and put on a pallet – all individual boxes with a slide entrance – and they all get a ripe queen cell from the incubator that's due to emerge later that day or the following day. The quad boxes also get given a queen cell, unless they have a queen, obviously, and then they get left to build up again. Then we'll split them again, into more individual boxes, and this is how it goes on.

Then we use split supers, that hold the same type of frame. They hold 24 frames in two rows of 12. These can be comb or foundation, and it goes on top of a single box hive or between the brood boxes of a double brood hive, so the queen can lay up all of the little combs. Then when they have brood in we take them off, shake the bees back into the hive, then we put it on top of the colony over a queen excluder, and leave them for an hour. The nurse bees go back up to the brood but you get no drones and no queen.

Then we take this super with its 24 combs of brood and put it on a hive floor, then we add another and another from other hives until we have five or six stacked up. We put a screen on top, strap them up and put them in the back of the truck. We take them to another apiary or the corner of a field, put them down and put a roof on and open up the entrance, and leave them. Then the next day we go back with the individual boxes. We split them all down. we start at the top of the stack and we go all the way down until we've filled up all our individual boxes, and all of our little combs have been used up. We then take those to another apiary and put them all out on pallets. They get a queen cell from the incubator. As the queens mate we remove them and they get another queen cell.

At the end of the season those individual boxes, which are often empty as the queen was sold, are piled up and then all the bees are shaken into an empty hive. We give it a mated queen and feed syrup. The empty half-combs go back into the split supers, 24 frames per box, all with brood in, and they go back on top of other colonies above an excluder. All the brood emerges and boosts up that colony. Sometimes we'll put three supers of 24 combs on one hive.

Steve: So then you've got just drawn comb left.

Peter: Yes, you've got it. We take the supers off, and the excluder, shake the bees in and the combs come back to my home yard and go into large blue barrels. They are fumigated in the barrel with sulphur, and they stay there until they are needed next year.

The mating nucs in the 3 frame boxes that still have queens in, which we are not going to use or sell get brought back to quad boxes. They go into the quad boxes, five frames plus one queen in each quarter, then we feed them with fondant or syrup. This is in the autumn, and the quad boxes go on top of strong colonies, so the warmth of the colony below goes up through a mesh floor into the quad box. At the end of November, we add wood shavings on top, which is the insulation. Once a month we just change the feed jars, which have fondant in them. We take the vacuum cleaner along, suck out the shavings, change the feed jar and then shake the shavings back in on top.

The next year the quad boxes get split again into three frame boxes and off it goes again for the new season. They've always got queens in them, Steve. The quad boxes are occupied 24/7, all year round.

Steve: Yeah, it's amazing isn't it?

Peter: We've always got queens. We overwinter about 100 queens in quad boxes but could easily do 500; it's just that I don't need such high numbers.

Sometimes if I'm wandering around an apiary in winter and I notice a lot of drone activity I'll make a note of it and come back later on. I'll take off the crown board and if they still have plenty of worker bees, so she hasn't been a drone layer for too long, I'll squash the queen and put a new one in from one of my quad boxes. I just put her straight in, no cage or anything. I introduce a lot of queens that way, I did about 200 last year, no caging, just put the queen straight in. I observe her for a couple of minutes and I have a water spray there to mist them if they get lively – towards the queen that is – and I only lose about one in every hundred. I'd never do that with a queen that had been sent in the mail, this is a healthy queen that has come straight from a mating nuc, and within 20 minutes of being removed from the nuc she's into another colony. I would never do it even with a queen that has been overnight in a cage, because she's no longer a laying queen.

Wood or poly hives?

Steve: No. Those shavings that you pack above the quad boxes...

Peter: Cedar wood shavings generally.

Steve: Yeah, that's just for your quad boxes?

Peter: Yes.

Steve: So, what about a normal production colony?

Peter: Just Kingspan or equivalent in the roof. All of our hive rooves are insulated, all year round. The bees love it. I don't like poly hives, personally.

Steve: No? Well you are making wood hives so I guess it makes sense to use them.

Peter: Andrew has some. They are warm, certainly, but I tried them out myself for a couple of seasons and I really disliked them.

Steve: Because?

Peter: I just find them clumsy horrible weak things. I think I'm too hard on hives, I don't find them to be robust enough for me. I don't like the way they wear. I like the poly nuc boxes; I think they're fantastic, but for hives, for honey production I absolutely hate the poly ones. They are bulky and clumsy. Wooden hives stack much better in a truck. Those poly hives with their stupid bulky floors take up too much space, I hate them.

The cycle of beekeeping

Steve: About the annual cycle of beekeeping through the seasons…you've talked about the queen rearing part.

Peter: Well, in the springtime we go around checking for food, obviously, and usually in mid to late March I put drone combs into all the ones I want drones from for inseminating. I also put drone combs into sister queens that are going to our isolated mating station.

Steve: So, for example, right now [Feb] would you already know which colonies they are?

Peter: Yes. They'll get drone combs because I want drones early. Last year the earliest queens I inseminated were just at the end of March.

Steve: Wow!

Peter: I'd found a couple of drone layers in February and had taken the queens from the quad boxes to replace them, and they went on to be good colonies. But they were full of drones, so I had mature drones early. I actually had three of those colonies but in one of them the bees chucked all of the drones out shortly after introducing the new queen. Of the other two, one of them had been an isolated mated queen but she was old and had just failed. She was a Dartmoor mated queen. The quad boxes I used to replace the drone layers were ordinary F1 open mated queens, but they had brood so I just left them to it. They raised their own queens with no help from me, because they were strong nucs with brood, but those queens would never get mated naturally because it was too early. So I inseminated the two virgin queens in the quad boxes with sperm from the mature drones from the drone laying queen. I've still got them now, and they were inseminated at the end of last March.

Steve: Incredible

Peter: It wasn't planned but that's how it happened. They are a pair of good queens; one of them is exceptionally good. I won't breed from her or anything, she's not for breeding. It was just to save the queens rather than kill the virgins and have the quad boxes empty.

A lot of commercial beekeepers wouldn't mess about doing that but it was at the time of year when I had time to do it, and it's good practise early in the season. It didn't really matter whether it worked or not, but as it happens it worked perfectly.

We start grafting every year on the 13th and 14th of April. That's every year, whether it's raining, snow or shine. Some of these will be inseminated – I know I can do that – but the majority are raised to be open mated. I pray for good weather in the first week of May because by the time those grafted queens are sexually mature it's the first week of May and that's when I want mature drones around and temperatures up around 18°C, preferably higher. I have an apiary that's in a little heat pocket so that's where I put the first lot of virgin queens; it's a sun trap and they always mate there before anywhere else.

Steve: When you're grafting, to guarantee that you have larvae of the right age from the right queen, is there anything special you do?

Peter: Well, I go to the breeder queens and put a brand-new drawn comb in four to five days before I want to graft. When I go back to it there will be lots of eggs and some tiny larvae, because she doesn't go straight in there to lay, the bees have to clean the comb up first. She might not lay in it for the first day and a half. That comb is marked so all I have to do is take it out, look to see if the queen's on it, and if she is I take her off, then I shake the bees off, wrap it in a towel and take it away to the apiary where I'm doing the grafting. I put that frame into a box above an excluder because I know that the following day I can do more grafts from it, and the day after that I can do another load of grafts from it because the eggs will have become young larvae.

Steve: I see, that's clever.

Peter: I'll have several cell builders and I put 20 grafts in each, plus I'll graft a few loose ones and just rest them down between the bars. Then the following morning I go round to the cell builders and take the cell bars out and see how many have taken. Any duds get taken off and replaced with good ones from either the loose cups or from the next cell builder colony as I move around them. I make sure I have 20 good ones in each colony, until I get to the cell builder and that may only be left with five. Then I'll go back to my frame of larvae, do another load of grafts to put in there, plus some surplus, so they end up with a full 20 as well. That's how I do it. I write the date of grafting and the date they are going to emerge in my diary. I know that 16 days later, from the egg, I'm going to have a queen, so from grafting day it's 12 days.

Steve: Where do you get the drawn comb to put in with the breeder? Did you say freshly drawn comb?

Peter: Just white comb, new comb. They haven't got to be freshly drawn. We are always getting them. You make up nucs and you put new foundation in, and you put foundation into a hive to replace other combs, double brood chambering hives – we always get plenty of drawn comb. I use deep boxes for supers as well and I always keep back some white combs to use for grafting.

Steve: OK

Peter: With the paint brush method of grafting that I use, a nice white comb is easier. With the old dark comb when you try to break down the cell walls you get all the pupa casings popping up and you can't see the larvae; it's horrible. I like nice new white wax comb. You can get rid of the cell walls easily leaving the larvae exposed so they are easy to pick up. It's much quicker.

Steve: Right

Peter: I think probably if you were using a steel grafting tool it would be easier with the dark comb because you can put the tool under the bottom and lift them out, but I don't use that method. Everybody that does grafting gets used to a certain way and has their favourite way of working. Some people like to use a cocktail stick, it's just whatever suits you.

We have talked about swarm control. I don't know if I mentioned it but we also split vertically using split boards as well. We don't need to be highly migratory, we move them in the spring to their summer apiaries, and those that are not near to heather will get moved to heather later on then back to the valley afterwards. We don't go in for pollination or taking our bees hundreds of miles away, we don't need to. We've got everything we need right here.

We spread them about because we don't want too many hives in one place. None of them do well if you've got it too overpopulated, so it's just a matter of spreading them out, within reason, within this area (15 to 20 mile radius). We have everything growing round here that we need. It's not arable land so we haven't got field after field of oilseed rape. If we want that, which I haven't for the last six years, we can move bees to the oilseed rape. The reason I haven't been doing it is that the weather in spring has been terrible most seasons and the oilseed rape is flowering earlier and earlier, because of the types they are sowing, so it's often in flower in the middle of March. The bee colonies aren't big enough to be collecting much from it so it's a waste of time, and the weather's abysmal.

Years ago, with the old types of rape it used to flower in mid-April until the end of May and it seemed to flower for a longer period, and the bees were strong enough by then to take advantage of it and make a lot of honey. It seems to have got earlier and the bees just aren't up to doing it. It's a lot of work to be disappointed. You're getting the odd good day when they bring a lot in, it's granulating, it's a pain in the backside and I'd rather be doing without it. I'd rather concentrate on the main spring flows like the sycamore trees and dandelions, and do other things like the queen rearing. Plus, some of our customers specifically do not want oilseed rape honey. They want our wildflower honey.

Selling honey

Steve: Is your honey mostly sold in bulk, in barrels?

Peter: Yeah, barrels and buckets. We've just had an email today, since you've been here asking how much summer honey we are likely to have to sell him, and the same for the heather. They say that the summer honey in particular sells very well. Last year we sold them three tons of summer honey and half a ton of heather.

Steve: It's an impossible question isn't it?

Peter: It is an impossible question. But as I say, we've got quite a few customers and we try to sort out a few tons each for them. That's how we try to do it. We can sell it, if we want to, all to one place. We've got one customer that will take all the honey we can produce, they'll take the lot, but it's too many eggs in one basket. I'd sooner spread it out a bit so that if something bad happens to one company I'm not going down with them.

Steve: You were talking earlier about how you use a lot of intuition when selecting queens, presumably the qualities that you are looking for are fairly standard?

Peter: Yes, I'm looking for bees that are low swarming, steady on the combs – I hate drippy bees – black bees are renowned for it.

Steve: So you'll be trying to breed that out?

Peter: I like to take a comb out and see that the bees are steady walking around on the comb, not flying up in your face, or running around like headless chickens. Sometimes you get hand stingers; I don't like those, I don't like stingy bees at all. Bees that you can work with are gentle and steady, low swarming and productive. I take particular note of colonies that are foraging in cooler weather, when the other colonies are quiet. That sort of thing I take note of. If that colony is regularly doing that, when the other bees are quiet and they are out working, and they are not robbing of course, then if they've got the other qualities that's one to breed from.

Steve: That observation thing…it might not look like you are very busy, just sat there watching bees, but that's part of the job, isn't it?

Peter: Yes, it's very important if you are breeding bees. It's not important if you are just producing honey.

Steve: So, are there times when you go to an apiary just to observe?

Peter: Yes, I'm around my bees all of the time, Steve. Aren't I Sandra?

Sandra: You certainly are

Steve: Are you glad to get rid of him [laughing]

Sandra: I don't see much of him in the summer, to be honest!

Peter: I'm out with the bees all of the time. I'll drive ten miles just to walk through an apiary to watch the bees, just to watch them, not open a single hive. I just go there to look at them. Sometimes when it's hot at 8 o'clock in the evening I suddenly take off to look at bees. I've regularly been putting queen cells in at 2 o'clock in the morning. I spend nights and nights up with queens emerging in the incubator, caging and feeding them. It's like having babies. I'll go out and get attendants if I think they are going to be caged for too long, so I'm out at 2am catching bees out of a hive, or mixing up little bits of pollen with honey and putting in caps, and feeding queens, putting damp bits of tissue over the cage – I treat them all like babies. While everybody is fast asleep I'm up all night dealing with that sort of thing, but I thoroughly enjoy it. I'm not doing it as a chore; I'm doing it because I love doing it. I want those queens to be good. It's probably a kind of madness [laughs]

Steve: Yeah, it probably is! It's an obsession, isn't it?

Peter: It is an obsession, yes.

Steve: I'm an obsessive person. The best way to learn about something is to become obsessed with it.

Sandra: He's obsessed with it, definitely

Bee diseases and how things have changed

Peter: The main change is to do with viruses, which is probably linked to varroa. There's been Nosema cerennae, Chronic Bee Paralysis Virus, Deformed Wing Virus…you didn't have these worries all those years ago. You had accarine and Nosema, and maybe a bit of chalk brood, and that was it. Now it seems there's lots of other viruses. CBPV is one of the big ones, there's a real problem with it throughout the country, and that's why Giles Budge and Professor David Evans have got the funding now at St Andrews, through the NBU, to do research on it. I've got one of the testing kits which they sent to me, so I can send any bees that I have with CBPV to them. It's been wiping out colonies big time. Some people have had 40 or 50 colonies in an apiary wiped out with it. It's been wiping out 50% of some bee farmers' stocks.

Steve: I thought it was because they had too heavy a varroa load.

Peter: No, CBPV is not varroa related. They think now that this was what acarine was in the past, not the tracheal mite but the virus. Over the last 4–5 years it's been getting worse and worse. It makes no sense. I had an apiary last year with 35 hives and one colony got it. It comes suddenly – at first you think you've got a spray poisoning incident, you've got dead bees all round the front, but not over a big area, just under the entrance. They can be 2–3 inches deep, with dead bees, yet a week earlier there wasn't a problem. All of a sudden you have piles of dead bees and bees trembling, shaking. They aren't black and shiny or anything, just normal bees, but shaking.

Steve: What about the black shiny ones – is that a different virus?

Peter: It's a paralysis, but this one they don't have to be shiny black. You will get hairless bees in there. They look shiny because they are hairless, the other bees have been nibbling at them, but they are not black, just shiny. The queen might be laying like mad, and they'll supercede her, and then she'll die, and that's the end of them, unless you re-queen, but then she'll die. Or the virgin will mate and she'll be superseded again. They keep dying and dying. You'd have to feed them. You can have a hive in the middle of an apiary that gets it – it goes on for 3 months, stinks to high heaven – you have to spade away the dead bees every week, they dwindle down to half the size, then all of a sudden it stops! No more dead bees, just like it's been switched off. They start collecting honey again, they build up and if you're lucky they may get strong enough to over winter, and they may never get it again. You don't see a sign of it in that colony again ever.

In another colony it goes on and on and on, and then they die out completely. I had an apiary with 16 hives and it started in the strongest one, and they dwindled and dwindled, and all 16 hives got it – every single hive. This went on for three months. All the hives survived apart from that first one. I moved that miles away to somewhere remote and they died. One of the colonies went down to 2 frames of bees which I put in a nuc with a virgin, and I moved them away but they lived, and I've still got them. All the other hives got over it, the other 14. It stopped like the day it started, and most of them produced quite a bit of heather honey, strangely.

Why is it that you can have one hive get it out of 35 and none of the others get it, even a hive six inches away, and then in another apiary they all get it – every hive. It makes no sense.

Steve: Is that the one where there are different strains, like an A strain and a B strain?

Peter: Yes, that's correct. You get the shiny shaky bees, they get over it quicker. In the other type the bees don't look any different whatsoever, they are not shiny, just lethargic, and they just fall out of the entrance and lie on the ground with their legs in the air and they just die. That's the one that normally kills the colony. Sometimes they get a touch of both of them and generally get over it.

Professor David Evan's work on Chronic Bee Paralysis Virus

Professor David Evans is a virologist who worked at the University of Warwick, and I got to know him because we were making hives for them for five years or so. They were getting bees from Colonsay, black bees from Andrew Abrahams, and they were doing research on them into varroa related viruses, comparing bees that had varroa with others that had never had it. David has moved now to St Andrews in Scotland. He's also a beekeeper and has lots of his own hives. David and Giles Budge have now got funding to research the cause and transfer, and hopefully the cure, for Chronic Bee Paralysis Virus. They haven't told me how much, but the word is "substantial funding" to do something like five years of research. They've sent out the kits to people that they have dealt with previously.

It started off with me phoning Giles and David to discuss whether the virus can be transmitted via the comb. I wanted to find out if I could use the comb from colonies that had had the virus, or died from it, in another (healthy) colony. We tried these ultra violet wands that you can wave over them and I don't know if that did anything but in the end I came to the conclusion that you don't need to do anything. If the bees have got the virus, and they get over it, and they are all on the same combs then it's obviously not in the combs, is it? I also noticed that putting extracted super combs from colonies that had the virus on other colonies had no ill effect, they didn't get the virus. It's totally different with EFB or AFB, obviously, but with this virus I don't think it's in the combs.

I read up a lot on how long a virus can live outside of its host, because I thought they would die straight away, but apparently not. They can live on work surfaces and so on for some time outside of the host. Bees with CBPV do a lot of pooing inside the hive and the virus is transmitted through bee poo. Anyway, I worked out that it's not transmitted enough on a comb to cause a problem, so the virus can't live that long outside of the bee. The reason these guys have got the funding is because there is so little known about it. It's such a serious problem but there is very little known about it, so thank the Lord for people like Professor David Evans and Giles Budge. I think it's fantastic that they've got the funding.

Steve: Yeah, definitely.

Peter: It makes you feel helpless, a problem like this, there's nothing you can do. We can get rid of varroa mites, we know what to do with those, Steve, but when you've got something that's invisible and your bees are dying, and there's nothing you can do about it, you feel terrible. We have to send them samples; some recently dead, some older ones, some samples of bee poo, you name it. We have all these little tubes to put it in and have to label it properly with dates, and complete a massive questionnaire. It's all to help with this research.

Personal and political bee stuff

Steve: What did you do before beekeeping?

Peter: I was a gillie in Scotland when I left school.

Steve: Was that to do with salmon fishing?

Peter: It was deer shooting mainly, and grouse. That was my first job.

Steve: That's not a normal job is it?

Peter: No, I suppose not.

Steve: You were odd from the very beginning [laughs]

Peter: It seemed normal to me. I always had bees though. Early on I went into the motor trade because I was interested in cars and motorbikes, but even then, I used to go off to the bee meetings, even when I didn't have bees. I still enjoyed going to the bee meetings, you know, the Association meetings.

Steve: How old were you when your Dad passed away?

Peter: I was about eleven years old. He was 32 and I was 11, and a year later I caught my first swarm of bees. That's probably why I got more involved with my Great Uncles, because Father had died. I suppose it had an effect on me. I never really got over it, probably never will. It happens when your parents die doesn't it? You think of them every day, I do with my Father even to this day.

Steve: Do you have a "bee in your bonnet" about some beekeeping related issue; something that vexes you?

Peter: No, nothing really. I'm not a zealot, I don't care what type of bees people keep, I don't care whether they are top bar hive users, natural beekeepers, treaters, not treaters – I couldn't care less. I just do what I do and as long as nobody interferes with that then I'm quite happy.

Steve: Do you think the Government is supportive of beekeeping?

Peter: Nope. Not at all. I don't particularly want them to be either. What I like to do with the Government is keep as far away from them as possible, and the less they have to do with me the better I like it. I don't want them to be supportive or help me or anything. If they ignore me I'll ignore them.

Steve: What about Brexit? Are you affected by that in any way?

Peter: I couldn't care less either way. It's a lot of politicians. As long as they don't affect how the heather's flowering or come down here messing up my fields of clover, or the forage for the bees, which I'm sure they won't, then they can carry on doing what they like. I think there are good and bad points to Brexit. I like the idea of our country being independent of Brussels – the laws they make there – but me caring won't make any difference so I'm not going to lose any sleep over it.

I'd like to see James Dyson as Prime Minister one day.

Steve: My Dad likes him

Peter: I think James Dyson would be an excellent Prime Minister. I saw one of his lectures a few years back and it was one of the most sensible lectures I've ever seen, about what's going wrong with this country, and trade, and the young people and apprenticeships, and building places to train young people. He made a lot of sense and at the time he was trying to open up Universities and training places, and the Council were giving him nothing. So, he moved his manufacturing to Taiwan! That could have all been in this country. All he keeps here now is the brains to his operation, his design team, and all the manufacturing is now overseas. He did that because his local Council were so blind to the employment he was going to create, and he tried his hardest to do it, and they just kept turning the man down. They were short sited really.

Steve: OK, how about this: if you could wave a magic wand to solve anything in beekeeping, a free wish, what would it be?

Peter: Cure Chronic Bee Paralysis Virus. That's higher on the list than varroa, for me. CBPV is getting worse and it's a worldwide problem which we don't know enough about.

Treating bees for varroa

Steve: Do you vary your treatments for varroa?

Peter: Yes. We monitor the percentage of infestation by doing sugar roll testing. The boys do that. We use Kilner jars and do a standard sugar roll test, into white plastic buckets, and we try to keep the infestation rate to 2% or below. You test about 300 bees so you want the number of mites to be below 6. We use various treatments from thymol-based treatments, which we haven't used for several years, Apivar (which is Amitraz) which is a very effective plastic strip-based treatment. Many years ago, there was Bayvarol and Apistan but it's a waste of space here as the mites are still resistant to it even after 12 years. That resistance, here anyway, is long lasting.

I use sublimation of oxalic acid, which treats the hive in 25 seconds. It's called Sublimox and it works off the generator which we carry in the truck. You don't have to open the hive, you just have a small hole. I don't use oxalic acid trickle. I'll never use it again.

Steve: You had problems with it?

Peter: I had big problems with it. I used it in 2006 and 2007. My average winter losses were anything from 0% to 7%, often about 2%, which is very low. When I used the oxalic acid trickle method it knocked them for six. This is 3.2% oxalic acid used at 5 ml per seam of bees. I noticed that the colony cluster numbers had gone down drastically about three weeks after treating and there was dwindling in the spring, and the colonies that survived were way behind any that weren't treated, and the losses were 35%! All the years before that we had lost 0% – 7%. I put it down to just being a bad year because this oxalic acid treatment was the bees knees, so I repeated the treatment in 2007 on hundreds of colonies, and got the same result. Pretty much exactly 35% losses. I've never used it since. The following year after that I think I had zero losses and since then it's never been higher that 7%.

I don't treat in the winter at all now, with anything. I make sure the treatments in September and early October are effective.

Steve: You still have supers on then don't you? Aren't they still on the heather?

Peter: No, the supers come off in the first week of September.

Steve: Right, OK.

Peter: The bees that aren't on the heather and nucs get treated in late August. Those on the heather get treated after being de-supered in early September using Apivar. As we are de-supering we go around putting the strips in. With sublimation we give them three treatments through October and they have had the honey taken off by then.

Steve: Would you do sublimation as well as the strips?

Peter: No, I might only use one type of treatment per season. Last year it was Apivar, the year before it was Apivar, but the two years previously it was Sublimox and no Apivar. They get three treatments, maybe four, depending on the mite infestation. I make sure that treatment is really effective. We do the sugar roll test before, and a week after treating, to make sure they've died. Here the ivy flowers in late September and through October, and it's quite common for the bees to be broodless in the last week of August and early September. You look through the hives and you think they are queenless, but sometimes there are two queens in there, from supercedure. Normally I see the old queen and she's just stopped laying, which is ideal for Sublimox.

I don't know how it is in other areas, but for us it's difficult to make nucs in late August because there is so little brood about. That brood brake is great for treatment because the mites are phoretic, but from late September it's like a switch goes on, the ivy comes into flower and they lay up right through October and into November. That's all the young winter bees, which we want to be healthy. Because we treat late, and feed late, there's no need to open them up in December or January to treat again. When we check them again in April they are perfectly OK.

Our quick detection method, which we use before a sugar shake – we don't just go around sugar shaking everything – is to take a table fork with us, and when we find a patch of drone comb we'll hook it out and see if there's any mites on it. We'll do that in several places in the brood box, because you can sometimes have clusters of mites, and if we start seeing 3 or 4 mites on drone larvae then we go to the next level and do a sugar shake. We take a measure of bees into the container, close the hive up, add icing sugar, put the lid on, shake them up and down, put them on the hive roof and carry on to the next colony. We come back round and shake the bees in the jars for a minute then take the lid off and shake the sugar through the mesh in the jar into a white bucket. We put a little water into the bucket so the sugar dissolves and the mites float, then we count them. If the infestation is above 2% then we treat them. If it's low we write on the crown board "SS" (for Sugar Shake) and the date, and the number of mites counted. We might come around two weeks later to check them again just to make sure.

Even though I make and sell open mesh floors the majority of my hives have solid floors. We might have two or three in the apiary with mesh floors but the rest are solid. I don't like open mesh floors, I detest them. The few mesh floors we do use, which are rejects from the workshop which we can't sell, are used to show that the treatments are effective. We put lard on the inspection boards, so that any mites that drop don't get carried away by ants and earwigs, and after treating we can count the mites. As far as I'm concerned if the treatment is working on one hive it's working on all of them, so I don't need to put a mesh floor on every hive.

Steve: Why do you detest open mesh floors?

Peter: I just find that bees over-winter better here with the solid floors. They were tried out first back in the 1920s, they are not new. I haven't had this problem, but some people say that mesh floors are good for preventing damp combs in hives. I don't get damp combs with solid floors. I did a test on some floors about ten years ago, just a primitive test comparing temperatures inside the hives with different floors using a hair dryer to simulate the wind, and the temperatures in the hives with mesh floors were much lower. So I think that over winter, in the wind, with a mesh floor the hive will be colder. People will probably say that's nonsense but to my mind it's not, so I stick to my solid floors, and it's what I have found works best for me.

I've noticed from the beekeeping forum that more people are coming back to solid floors and also people are leaving the inspection boards in over winter, or just having a small area of mesh floor open for ventilation rather than the whole thing being open. I've noticed that with solid floors the bees will cluster right down to the bottom of the frames and onto the floor. With an open mesh floor, they are up high, which tells me that they are moving away from the cold.

Steve: Yeah, it makes sense.

Peter: Brother Adam tried them out back in the 1920s. All I heard was that he wasn't impressed by them, and he never used them again. But back then it wasn't about varroa, it was more the mouldy comb business, but if they were any good he would have kept on using them, because, like Manley he was way ahead of his time.

In beekeeping everything seems to go in a circle, Steve. They will come up with a new hive design, but when you look back you'll find somewhere that they were doing this 100 years ago. It's like re-inventing the wheel over and over. The reason things are being done the way they are is because they found out years ago what works and what doesn't.

Educating yourself on beekeeping

Steve: Do you read much about beekeeping?

Peter: I read lots of research papers, which I get from the internet. I'm really interested in any scientific peer reviewed research papers about bees, especially about bee diseases, but anything to do with bees. I'll read it for hours and hours. Some of it I believe and some I don't, but I enjoy reading it.

Steve: What about beekeeping books for beginners, any that you recommend?

Peter: I haven't read any for years but I would have thought Bees at the Bottom of the Garden by Alan Campion would be good, and that new Haynes Bee Manual, and Ted Hooper's book, Guide to Bees and Honey. My favourite books are Brother Adam's books, Beekeeping at Buckfast Abbey and Bee Breeding; I like Queen Breeding and Genetics by Eigil Holm although the translation isn't great; Background to Bee Breeding by John Atkinson although it's not for beginners – sometimes you have to read the same page 5 times to get what he's talking about, but my favourite of all time is Honey Farming by Manley. I've read lots of others, some of them are nothing more than rubbish, they are telling beginners rubbish, and some is just regurgitated rubbish. I don't advise beekeepers to watch the internet too much for information, especially not You Tube, because 99% of that is garbage. Complete rubbish made by numpties. 1% of it is very good, people like Mike Palmer and others of his ilk.

To checkerboard or not to checkerboard?

Steve: Do you remember Walt Wright and, what's it called, checkerboarding?

Peter: Ask Mike Palmer what he thinks about that!

Steve: I don't think I did ask him, actually.

Peter: I'll tell you what he'll say; it's the biggest load of nonsense he's ever heard, utter garbage. It stops bees swarming automatically, you'll never get a swarm! I remember him writing about it, Walt Wright, on Beesource. There were days and days and days of him and Mike arguing. Oh yeah, they've had arguments on the forum. There are threads on there that go on for hundreds of posts. I've read most of it and I've got the same opinion on checkerboarding as Mike. I think in our climate, with the unpredictable weather, it's not such a clever thing to be doing. Maybe if you live in the States where they have almost got a guaranteed climate, more of a continental climate rather than our Maritime one, it might work better. Here it can be cold one day, a heatwave the next day and then snowing the next, so I think it would be dodgy to do it. It's OK if you are just talking about honey combs, taking away the arc of honey from above the bees, like taking some combs out and putting empty ones in, that's fine.

The problem with shallow boxes like Nationals is they also tend to fill it with pollen, which is a complete pain if you keep under supering rather than top supering, because you end up with every super having a fair proportion of pollen in it. It proves that the National brood box is probably a bit too small for a prolific bee, and you don't get the problem with double brood boxes. We use about 50% doubles and 50% single brood boxes and I like to leave that super above the brood box where it is and add more supers on top. It doesn't make any difference from a honey yield perspective; there's been research done into that and it makes no difference whether you top or bottom super.

Steve: What about the honey arc on a brood frame? I read somewhere that the queen won't want to cross that so I wonder if I should break up that honey arc?

Peter: Generally, Steve, if they have a honey arc on the top of the brood frame they probably have a box of about the right size and you just super above it.

Steve: If they see there's honey above their heads how will they find out that there's more space above that for storing more?

Peter: They find out, don't worry about that. If you had a standard National brood box you often won't find a honey arc because it's wall to wall brood, every cell right to the outside.

Steve: Right.

Peter: When they get to that stage, if you've got a nice prolific queen and she's laying up every cell, there's only one thing you need to give them – another brood box on top. Not a super; a brood chamber. If she's a less prolific queen and you have got the honey arc then you don't need another brood box because she isn't going to go up there, you need to put shallows up there. Even when they are five supers high they find the space; they are up there filling the top box. With top supering you end up with lovely white combs up there, so that's where we get our cut comb. The ones lower down have all of their little boot prints all over them and they get dirty, and there's no pollen in the top ones either. That's why we like top supering. We do a lot of cut comb or whole combs being sold intact.

Providing space and swarm prevention

Steve: Regarding the provision of space, say it's spring and the colony is growing, and at some point you think they need more room, when do you know the right time to do it?

Peter: If she's laid up ten frames of an eleven frame box or she's laid up all eleven then she needs another brood box.

Steve: You wait until that late?

Peter: Yes, sure, she'll get another brood box then. We take out one frame of brood from the bottom box and put an empty comb back down there and the one frame of brood goes in the middle of the top box. The bees are up there instantly and will draw out any foundation next to the brood and she'll move up and lay in those frames.

Sometimes if it gets to just before that stage and it looks like they might be going into swarming mode we'll do a Demaree. We take another box of combs, empty combs, we'll find the queen and she'll go in the bottom in a new box of empty combs – just the queen, no brood, we just drop the queen in there. Then the queen excluder goes on top of that box, then some supers, and then we'll put all of the brood up top, with a small top entrance for drones. The queen now has unlimited laying space, all of the nurse bees thinking of making queen cells are at the top out of the way, and on the next inspection we'll find queen cells in the top box. We just shake the bees off and break off the emergency cells. Over time, as the brood emerges in the top box we remove the empty comb and swap it with brood combs from the bottom box, including eggs and larvae. You are always giving the queen plenty of space to keep laying, keep laying, keep laying and you remove congestion by having the young bees in the top looking after that brood.

That's called Demareeing, and they work like mad filling up the supers. At the end of the season all of the brood emerges from the top box and then they fill it with honey. You then have a choice; either put the top box back on the bottom box which gives them their honey for the winter, or if you need the honey you can extract it and put empty combs on top then feed them.

Steve: Let's say you've got a single brood box, at what point do you start supering? Say the queen is not going to lay wall to wall and a single box is OK...

Peter: Seven to eight combs of brood.

Steve: Right.

Peter: You mean putting the first supers on in the spring?

Steve: Yes.

Peter: We super on seven to eight frames of brood. That can be anytime from late March to early April.

Steve: That will be drawn comb, won't it?

Peter: Yeah, it's always drawn comb to start with – sticky ones as well, because they were extracted last year. They are up there like a shot.

Steve: Then the next one?

Peter: Well as they expand it's just observation of how prolific the queen is.

Steve: Is there a rule of thumb like, say, when five of the super frames are capped or something?

Peter: We never go in there like some hobby beekeepers manipulating the combs in the supers. When it feels heavy we stick another one on. When it looks full of bees and it feels heavy or half heavy, another super goes on top, Steve, and that's it. Sometimes if it's a really good flow like on borage we'll stick two supers on, or put a brood box on to use as a super. Our extractor takes deep frames as well as shallow. The uncapping machine does eight frames per minute, both sides.

Types of bee hives

Steve: You're using Nationals but somebody using a bigger box, like a 14 x 12…

Peter: An abomination!

Steve: Yes, I don't like the 14 x 12's either.

Peter: If you want to use a bigger brood chamber I'd go for the Commercial rather than the 14 x 12, a standard British Commercial hive. I used to run those but they were too big for around this area. I gave them all to the bee inspector and within two years he got rid of them.

Steve: So if you have one of those bigger boxes you are saying that there's no need to go to double brood ever?

Peter: That's right. Those 14 x 12s are big horrible ungainly bee squashing things, sagging in hot weather, take an age to examine – they are a nightmare. I'd sooner have double brood Nationals, any day, than a big single chamber.

Steve: Would you call a Langstroth a big chamber?

Peter: No, not particularly, not a lot bigger than a National. It's a bit bigger but not hugely so. I like National double brood for the quick swarm inspection. It's not guaranteed, but on the percentages, it makes sense because you can inspect them in a minute or less just by looking for cells at the bottom of the frames in the top brood box. That's where the cells go, and if there are none you are done. If you see some cups with larvae in then we'll look through some middle combs to see what's going on, to see the laying rate. Then if we need to do anything else we can. If you go through 20 hives in 20 minutes, that's not bad, that's speedy.

But it's not fool-proof and you can have a cell somewhere else, probably a supercedure cell, and they decide to swarm on it instead. In our case the queens are clipped. One of the first jobs in spring is to go around finding any queens that aren't clipped and then they get clipped. I never clip them in their first season, apart from the inseminated ones. None of the naturally mated queens get clipped in their first season, but in late March to April the following season, when we are going around doing our first checks, we find the queens and clip any that aren't done. I do that because I prefer to lose a queen rather than the whole foraging force of that hive. I can easily replace a queen but I can't replace 50,000 bees very quickly, hanging up in a tree, which breaks my heart. I think about all that honey they are not going to make for me – I prefer to just lose the queen.

We don't like losing queens either, obviously, which is why we are very hands-on. You don't find many big commercial operations checking their hives every 8 days, but we do. We go through every single colony every 7 to 8 days, which is why we are doing it from daylight to dark. You can have a honey farmer with 1,000 hives and another one with 600 hives, or 400 hives, and he can produce just as much honey as the one with 1,000 hives because he goes through every colony every 8 days. The guy with 1,000 hives is sat in the pub talking while his bees are swarming left right and centre. The guy with the 400 hives is getting as much honey but he has to put in the work to do it.

Do you see what I mean on that?

Steve: Yes, I do.

Peter: That's how it works. We could have 2,000 hives and never go round and visit them. We could super them up in spring, go back and take them off in the autumn, but we wouldn't have any more honey than the hives we've got now that are intensively managed. I wouldn't enjoy doing the other way one little bit. I wouldn't call that beekeeping even.

Future Plans

Steve: My last question is, what are your plans for the future?

Peter: The Bahamas sound very nice! My plans are to carry on as long as I can, like most old beekeepers, breeding queens.

Steve: You'll probably always do that won't you?

Peter: I want to do that until I'm carried away in a wooden box. I'd be quite happy if I dropped dead one day out in the apiary. If I've got to go I'd rather go quick and be out doing what I enjoy doing, with my bees. I hope that my boys will take over and make a success of it in their time and probably turn it into a bigger business than I ever would. That will be down to them. I've shown them what I do, and they've spent enough years with the old man around, so it will be up to them. That will happen long before I pop my clogs though.

My ambition is to stay on as long as possible, keep breeding bees, and then one day drop dead in an apiary.

Steve: There are worse ways to go.

Peter: I've got no real desire to go anywhere or do anything else. I feel quite like I'm in Heaven right where I am. I would really like to go to New Zealand; I've always wanted to go there. I might go and visit Mike Palmer for a holiday and I'd love to go to New Zealand, because apparently it's very much like how England was in the 1950s, with a low population, not like here now.

Steve: I thought New Zealand was lovely. Thanks Peter, we're done now.

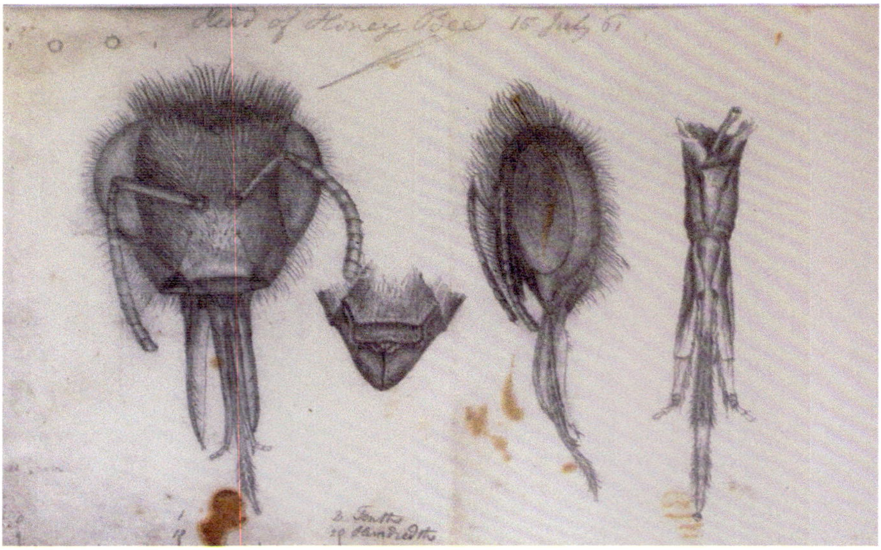

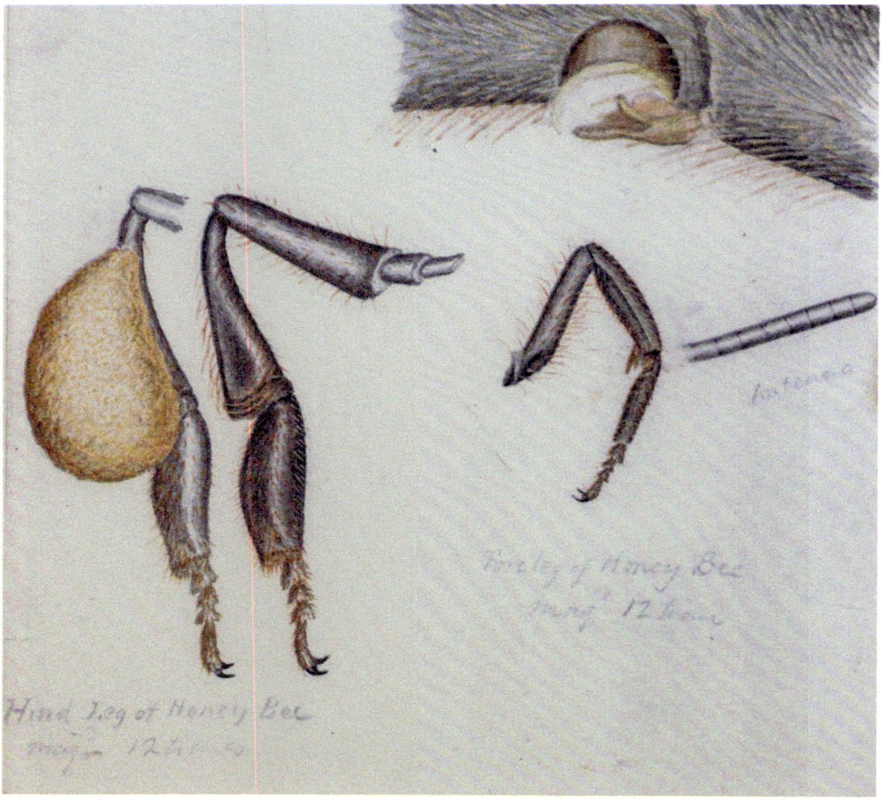

William Bray sketches

Peter Bray, Leeston, New Zealand

'Starvation and queenlessness are the biggest cause of hive deaths worldwide, followed by varroa and other things. Varroa is only a cause of hive death in about 5% of cases on average... I think of myself as being reasonably environmentally minded, and I think that the biggest cause of extinctions worldwide is taking land and habitat away from natural activities.'

Introduction to Peter

Even as the idea for writing this book was developing in my mind there was one thing about which I was certain; I was going to visit New Zealand. I love beekeeping and have learned so much from the people in this book, but one of my primary reasons for doing this was to have an excuse to travel. Given the considerable distance involved, I decided to fly in style and chose Emirates Business Class for my journey. My son Alex was delighted to accompany me, and I was glad to have a companion to keep me sane. Alex and I both play an online game called Hearthstone and have done so since it was in its beta version. We often joke about all of the places across the world that we have played this game. Other people may do site seeing, but often we sit in our hotel room playing games. Is that bad? I think it might be.

I found New Zealand to be a beautiful country, almost perfect. When we arrived in Christchurch in January 2018, which is mid summer for them, they were having a heatwave. It may be a cliche, but both Alex and I found the place to be stunningly beautiful and the people friendly and welcoming. It had a slight "Canada" vibe; both countries have the most staggering natural beauty with diverse and liberal-minded populations. We walked in the lovely Botanical Gardens most days and had brunch at various eateries dotted about the place. The quality of the food in the restaurants was top class, the sushi was heavenly, and I even developed a taste for smashed avocado on sourdough toast, washed down with a green smoothie.

The downside? The biggest one is earthquakes. A devastating quake hit Christchurch in February 2011 claiming 185 lives. It flattened a section of the city, and they were still busy putting it all back together again in 2018. I spoke to several of the locals about this, and they all seemed almost blasé about earthquakes; they are just part of life here. Then there is the hole in the ozone layer which means that you have to plaster sunscreen over exposed body parts before venturing outside unless you want to risk skin cancer. The other disadvantage for me, and perhaps the reason why I'm not planning on emigrating today, is how far away New Zealand is from my family and friends in the UK. I have seen paradise and chosen to live in Manchester; I must be mad.

Peter Bray owns and runs Airborne Honey which is the oldest honey producer in New Zealand. After I had arranged the interview and made travel plans, I was horrified to discover that he is no longer a beekeeper at all. It turned out not to be a hindrance. It was quite eye-opening to get a different perspective on the honey industry. There isn't much in New Zealand beekeeping that Peter doesn't know about, and he is forthright in his strongly expressed views.

Talking with Peter

How the company began

Steve: You were saying that you used to have 6,000 hives?

Peter: Yeah. Well, the company started with my Grandfather. A lot of this is on our website. He went to boarding school in Christchurch and one of the masters at the school got him interested in beekeeping. At home I've got a set of drawings that were done by my Great Great Grandfather, and they are really highly detailed drawings of honeybee mouth parts, and legs and things like that.

Steve: So he'd been using a microscope for that.

Peter: Yes. I also have the microscope, which is now an antique. I don't know whether or not those drawings influenced my Grandfather, or why my Great Great Grandfather was into that stuff. He was an engineer on the railways and worked for Robert Stephenson. He'd been involved in Egypt and then came to be the first engineer for Christchurch.

Steve: When was that?

Peter: He arrived here in the 1850s. The provincial government didn't have money so they paid him with land.

Steve: Right, sounds good.

Peter: He was incredibly wealthy and by the end of his life he wouldn't have been a billionaire but he'd be a multi-millionaire. His son was deaf and dumb, and also a little bit arrogant, and he squandered the lot [laughs]

Steve: Wow.

Peter: My Grandfather then grew up in poverty in a place not too far away from here. He got a scholarship to Christ's College which is the pre-eminent boarding school in the city, and he started to climb out of the mire. His interest in bees started there, and when he'd left school he biked around the South Island as an inspector for the apiculture department. They had a move to shift hives into removable frames, and later they had an Apiaries Act which said that all hives had to be all frame hives, so that they could inspect them for AFB. At the same time they introduced the law that all hives with AFB had to be burnt.

Steve: That hasn't changed, has it?

Peter: No. It means that there is no antibiotic feeding allowed, which I think is a good thing.

There are numerous cases in New Zealand where beekeepers have taken over completely diseased outfits and they've completely and utterly cleaned it up, and ended up with zero AFB after a period of time. It's pretty rigorous. One of our writers, Cliff Van Eaton wrote a book about the process of doing this called "Elimination of AFB without the use of Drugs". There are a number of companies that have done it and been successful with it.

We got to the point where my Father took over from my Grandfather and they developed the packing business. I didn't really want to join the business but I liked the idea of packing honey and processing it, because I'm more of a technically minded individual, rather than a green thumbed individual. My Grandfather and father both grew veggie gardens and orchards but its anathema to me.

Building the laboratory

Steve: Nice to eat, but you don't want to actually grow the things?

Peter: Yeah, pretty much. So, I ended up going to the Queensland Agricultural College, just West of Brisbane, and did a two-year associate diploma in beekeeping. I had a very good tutor there that was not only extremely good with bees but had quite an influence on some of the other faculties. I was introduced to laboratory techniques and things like pollen analysis. I got interested in that and saw that there was an opportunity to be able to differentiate honeys according to their source. When I came back into the company that became quite an interest for me. I started back in 1980 and by about 1983, I think, we had a laboratory, or at least a microscope and an area to do analysis.

Steve: The idea was to do pollen analysis, was it?

Peter: Yes, we saw a need because we were selling honey to Europe. We'd call it clover honey and they would say, "No, it's not!" They would say it's got this, this and this in it. I thought that maybe they might be right, or at least if they were not we need to know more about it so that we can speak from a position of knowledge rather than ignorance. We developed our laboratory and employed somebody and trained them. They were also trained by a local palynologist, just up the road here in Lincoln, who worked for what is now Landcare Research. He did a lot of research into fossils in things like coal, and pollen analysis is a fundamental process that they go through.

The range of analyses that we incorporate has grown and grown. Now we've got two HPLC machines...

Steve: What does that mean?

Peter: Well, HPLC means High Performance Liquid Chromatography. You take a sample and prepare it in a way that you can inject it into a loop in a high-pressure environment, so you're pumping that sample through a column. There are different columns packed with different materials. It's not unlike dipping a stick of chalk into some ink.

Steve: It separates out

Peter: Yes, it separates it into various components. With HPLC you are pumping sample through at thousands of pounds per square inch. You can take honey and put it through the right sort of column and it will separate the different sugars out. You get a base line and you get a peak coming out, then another and another, and each peak corresponds to a particular sugar. The area under the peak is proportional to the volume of that sugar. You need a range of different detectors to pick up certain things. There are mass spectrometers which work well because everything has mass, and some people use UV light because some chemical bonds absorb UV light. We use something called ELSD [Evaporative Light Scattering Detection].

Steve: There's some powerful stuff going on there...

Peter: Yeah. For us it's sugar analyses. In New Zealand one other company has just started, but we are essentially the only company doing sugar analysis of honey. Sugar is 80% of honey, and the sugar chemistry is highly complex. We can look at things from an identification perspective, or a nutritional aspect, or a performance aspect.

Steve: Are you working out the ratios of the different sugars in the honey?

Peter: Yes. There's a suite of things when you're looking at identification. We use the duck analogy, so if it's got webbed feet, flat bill and quacks, it's a duck. If we only know about the webbed feet it might be a seagull. One element on its own won't give you the full picture. The more elements that you choose, particularly if they are discriminating elements, the better idea you will have. Some honeys are not remarkably different from other honeys for their sugar ratios but some are very different.

Economics and the economy

Steve: I don't think you have oilseed rape here do you?

Peter: A little bit.

Steve: We never used to have that. It was clover all over the place, which you still have lots of don't you?

Peter: Yes

Steve: When I was interviewing David Kemp, who was assistant to Brother Adam many years ago, he was thinking back to those years and how much things have changed. I mean, oilseed rape is everywhere in certain parts of the country, and that has completely changed beekeeping. People carry their hives to the fields.

Peter: So, they'll be getting paid for pollination?

Steve: Well, there's some of that, in orchards and such like. We don't really have many huge commercial beekeepers in the UK like they do in America, and here I think.

Peter: There's a few here claiming upwards of 30,000 hives, and a few in the States with 90,000.

Steve: The biggest in the UK is called Denrosa, owned by Murray McGregor, who I met a few months ago. He's got 3,500 and that's literally the biggest. We have lots of hobby people like me with about ten hives.

Peter: We've gone through a change here. It pretty much happens everywhere in the western world. If you look at the data for the USA they used to have about 5 million beehives, and the numbers dropped pretty consistently until they had their colony collapse disorder, and then their numbers started going up again. You might consider that that's odd, and you might expect the colony collapse to make numbers go down, but the thing that makes their numbers go down is an economic issue. When beekeepers are profitable people think that it's great and they add some more hives.

Often people start out as a hobbyist, giving honey away to their friends, and then they start to sell some at the gate, and if that goes well they get more hives. If their colonies were struggling and they weren't able to sell much honey they would start to lose interest and maybe let them dwindle and not run so many hives. That's what it's like in a commercial business but on a greater scale. If the price of honey goes down the number of hives decreases, and if the price goes up the number of hives increases.

In the USA in the 1960s their per capita income was really low, but they had

The Lab –Airborne Honey

an economic transformation and their per capita income went higher and higher. The single biggest cost is labour, so to make it work they either had to have a radical increase in their income per hive or the hive numbers would decrease, which is what happened. What caused the increase in hive numbers was the money from almond pollination in California, so that now there's something like 1.8 million hives in the USA going to almond pollination out of a total of about 2.5 million hives across the whole country. 70% of the hives in the country all get trucked across to California, but they are getting \$150–\$180 per hive.

In New Zealand we had honey prices that were barely able to sustain the hive numbers, but then we had a boom in kiwi fruit. At one point around a third of the hives in the North Island would have been going to kiwi fruit pollination, and suddenly they could get \$100 per hive, and people were saying, 'Wow, that's incredible," but then it went to \$200 per hive. I'm not sure what they are paying now, but with the advent of Manuka, beekeepers are getting \$1,000 per hive or whatever it is, so they are not so interested in kiwi fruit pollination anymore. Those price increases have driven huge increases in hive

Bottling honey

numbers; the pollination plus the Manuka phenomenon. In 2000 we had around 320,000 hives I think, and it dropped a bit when varroa came along – for a number of older beekeepers it was just too much for them to face dealing with it so they exited, but numbers started to increase again and now we've got a million hives. It went from 320,000 to about 290,000 and then rose up to 1,000,000 hives. Our ten-year average is 32KG per hive whereas the 6 year average before varroa was about 30KG per hive, so productivity has actually increased as well.

Moving from bee-keeping to research

Steve: You're not a beekeeper anymore are you?

Peter: I can be a beekeeper. When we ran 6,000 hives I was a beekeeper. I'm not a good beekeeper, in terms of being able to organise my day around the bees, and I was being pulled in many different directions; trying to create markets and provide products to sell to those markets. We were selling ten different types of honey but our bees were only producing two types, so there was a lot of capital tied up in bees and not enough in the processing, packing and marketing side. We made the decision to sell the hives, which we did in the late 1980s and early 1990s and never looked back, basically. The site here had an extracting plant that had to be operating full time at this time of the year and the packing and marketing took a back seat. We used to be processing 600–800 tons a year and bringing in 100–200 from our bees, and we'd get all sorts of robbing issues with bees. Now you don't even see a bee around here!

Steve: It's interesting that you chose to go the way you did because the last guy I interviewed, Murray McGregor, took the decision to just do bulk honey. He focusses on the beekeeping and sells it in bulk so he doesn't have to worry about the packing and marketing. You've gone the other way. It seems that to do the whole thing is probably too much.

Peter: Well, some people do do that, and they do it well, but one of the issues with beekeeping is that corporate beekeepers are very rare. Yes, there are examples of them around the world, but there's a window in the spring when beekeepers can only keep so many hives, and they need to work daylight to dark for seven days a week during that window. There are not many corporate organisations that can employ labour that will do that, and the passion to do that comes from what some call the sado-masochism of beekeeping. You have to love it to do that, and to get employees to love it and do it is a really hard ask. There are a few profit sharing models that work but generally the large beekeeping operations are all about how many employees can you keep going under the guidance of a prime motivator of the business, and how many hives can you get round in that period. It will vary from location to location, from operation to operation, but there will be that constraint of "how many hives can I run per person?" That ultimately drives the amount of honey produced; the gross income for the given labour force, and therefore the profit you can make.

The Manuka phenomenon

Some organisations can do that and run a marketing arm as well, but very few can. We've got one in New Zealand called Arataki, who are a competitor of ours; we know them well and have been in business alongside them for years and we get on with them. They have a large number of family members in the business, and that's often the case when you get these large organisations; there's often multiple family members.

The Manuka phenomenon breaks that somewhat. We've suddenly got massive amounts of capital being invested in the industry fuelling lots of growth.

Something like 30% of the industry has been in it for more than three years, so 70% are basically newcomers. That's profound.

Steve: When did all that kick off then?

Peter: The Manuka?

Steve: Yes.

Peter: The genesis of it was when a researcher back in the 1980s found that honey was anti-bacterial and great for wound dressings. He was passionate about it and went forward with it, and the industry supported him through marketing initiatives and PR. That carried on through the 1990s so that by 2000 the price of manuka was starting to move up. It probably peaked in the last couple of years.

There are some issues arising out of radically increased hive numbers. We've gone from a crop of 8,000 to 9,000 tons average to 30,000 tons now.

Steve: Does that mean that you are having to plant more of the trees?

Peter: Not really. There are some people out there planting manuka. It's a get rich quick, goldrush mentality, and there are going to be a lot of people burnt really badly. The world market is \$3–\$4 per kilo whereas beekeepers in New Zealand have been getting \$10–\$12 per kilo for non-manuka honeys, because they are being blended into manuka. There have been around 9,500 tons of exports over the last few years at prices of \$30–\$40 per kilo on the world market. When the world market is \$3–\$4 per kilo it tells you that most of that is being sold as manuka, because people aren't buying clover honey from New Zealand. Ten times the global price!

You used to see Rowse importing clover honey into the UK and putting a "New Zealand Clover Honey" label on it, but that's died because they can't afford to pay the \$10–\$12 per kilo for clover honey. So where is the clover honey going? It's going into manuka.

Steve: Is it a bit like the Champagne in France thing, where there are probably more bottles sold labelled Champagne than it's possible to produce in that region?

Peter: I don't know, because it's hard to get statistics on that. There have been lots of claims made about that. One industry organisation said that the annual crop of manuka was about 1,700 tons and yet the export stats were showing 8,000 tons of honey being exported. It's not all manuka, but if you assume that 80% of it is, which is probably not far wrong given the prices, then you can see a discrepancy.

Steve: You actually have a special label on your honey giving an analysis of what's in it, so I guess your integrity is about as high as it gets?

Peter: Well that's not what the others would say. They go down a rather circuitous rabbit hole where they say that manuka is famous because it contains a special active ingredient, and without that it isn't manuka. Therefore, if your honey doesn't contain that active ingredient and you sell it as manuka then you are defrauding the consumer. That's not a bad argument, but the problem with it is that they have told lies about what it does! The consumers buy it because they believe that it's going to do something amazing for them when they eat it.

The reality is that there is no evidence at all that it has any benefit once you ingest it. You can find claims for digestive health, cure for irritable bowel syndrome, gastro reflux disease, stomach ulcers, stomach cancer, cancer in general and so on. The public lap that up. We get calls from people saying they are driving by and are going back to the UK soon and they have been told to pick up some of the best manuka for their relative who is dying of cancer, and which one do we recommend? You can't really help them.

Steve: You need something a bit more than honey for that!

Peter: Yeah. These are the myths that are out there.

Steve: It's very good on wounds though, isn't it?

Peter: Yes, it has benefits, but there are studies which compare standard wound care preparations to manuka. One study I'm aware of said you get about 80% success and 20% failure for both treatments. The thing with manuka is that it costs more and it hurts more, and it's more difficult to apply than standard wound care preparations, so it actually makes more sense to use the standard stuff. If you get a problem with that, then you are likely to apply manuka, as a second step.

They then tested it on peritoneal dialysis catheter sites, comparing manuka treatments to the standard kits, and again there was little difference. Not only was manuka no more effective, more expensive and hurt more, they also found that in diabetic patients they were about 35% worse off than with the standard preparations. It's to do with the Methylglyoxal (MGO) in the manuka. It's produced naturally in the body. You've got glycolysis, which is how your body breaks down glucose, and during the steps of that process there's a Dihydroxyacetone Phosphate (DHAP). You might know that Dihydroxyacetone is the active ingredient in the manuka nectar that then turns into MGO. Well, in glycolysis it's almost certain that DHAP turns into MGO, so we've got this low grade production of it in our bodies. We also have this enzyme system, called the glyoxalase system, to mop up MGO. Diabetes is a disease affecting glucose uptake in the cells so it possibly isn't a surprise that if you overload the glucose conversion process with extra MGO at the site of application of manuka, you get a decrease of the healing process.

So manuka becomes an increasingly small wonderment, if you like, the more you look at it.

Steve: Many of the beekeepers back home in the UK are jealous of the amazing marketing of manuka honey and how successful it's been. Most beekeepers think, "Yeah, it's just honey with a funny taste, what's the big deal?" I think they admire the success, but I don't think many beekeepers think manuka is anything particularly special.

Rise of the Leptospermums

Peter: Well the Leptospermums are special. There are 83 species of them in Australia and New Zealand, and there are lots of them in Australia that have got far more DHA in the nectar, and therefore potential for more MGO than any in New Zealand. Our averages are about 3,000 parts per million of DHA; they've got some over there that have got around 17,000 ppm, and they've got a much larger area. But, we got it first!

The word "Manuka" is a Polynesian word, and there is a Manuka reserve on the big island of Hawaii, a Manuka Street in Tonga, and you've also got Manuka Streets in Australia because 30–40 years ago when they named those streets everybody called the Leptospermums "Manuka". That was just the common name, but now the New Zealand organisations are trying to trademark it around the world, and they are saying it's a Maori word.

To me it's wrong because you need a naming system for honey plants around the world that is consistent and makes sense. You can use the common name or the botanical name, and where you use the botanical name you state the country of origin. That all makes sense, and you need a system like that which works seamlessly around the world. If you suddenly go, "Actually, you know what, clover is an English word, and only England can call clover honey 'clover honey'." To me it really doesn't work. We've got Viper's Bugloss here...

Steve: Is that what we call borage?

Peter: Well, exactly! We also call it borage. North of Christchurch, traditionally the local beekeepers would call it borage. It's Echium vulgare. Now we've also got Borago officinalis, which is culinary borage or blue borage. But, North of Christchurch Echium vulgare is called "Blue Borage" – there's even a

Blue Borage Apiary. South of Christchurch it gets called Viper's Bugloss. In the United States it's called Blue Weed. These are common names.

Several hundred years ago a character called Linnaeus created the double banger naming system, because scientists want precision about what they are communicating, whereas Joe Public wants easy names to use. The notion of a common name is what the general public use. We've got a raging argument here about two plants; Kunzea ericoides and Leptspermum scoparium being called "Manuka". Kunzea ericoides used to be Leptospermum ericoides. In the most recent edition of the Maori dictionary both plants are labelled manuka.

There's a little book published in 1978, written by the government honey grader, and published by the national beekeeping association (membership was compulsory if you had hives) and edited by the chief apicultural advisory officer. Back then the relevant government department was the Department of Agriculture and they administered all of the legislation pertaining to the beekeeping industry. In this book there is no listing for Kanuka, which is Kunzea ericoides, but there is a listing for tree manuka, and manuka, with lots of alternative names like white manuka, white tea tree, red tea tree and so on. The Maori also had a range of different names for them such as kahikatoa – "warrior wood" and makahikatoa – "white warrior wood". Once the marketing machine got going "manuka" could only be Leptospermum scoparium because it's the only one which produces the active ingredient, and therefore it must be manuka, which is a rather circular argument. That's the one that's taken hold. They have the same smell, the same taste, both are thixotropic, one sometimes has the active ingredient and the other one doesn't.

So, Leptospermun scoparium is the one true manuka, and sometimes it has activity and sometimes it doesn't. Just because it's that plant doesn't mean it has the active ingredient. If you are going to say, "I've got manuka honey," you also need to say what level of activity it has, if that's what your consumers are interested in. I see that as an easy step and it fits with what the rest of the world has agreed through the Codex Alimentarius Commission, when they created the codex honey standard.

Good years and bad years

Steve: OK, thanks for that. Most of my questions are related to beekeeping but I'll ask them anyway.

Peter: Fire away.

Steve: Can you think back to when you had your best and worst years, and say why that was so?

Peter: Well, we're linked at the hip with the beekeepers anyway. We had one absolutely disastrous year, and at that time we did have our own bees, and we didn't have enough honey to meet our markets. We were extracting willow honey and buying willow honey to use it to supplement our manufacturing markets. That would have been about 1983.

Steve: Was that bad weather or something?

Peter: Drought. In Canterbury here we get quite a significant drought effect. This year shaped up really well in the spring and it looked like it was going to be a good crop and then it just stopped raining. We had no rain in November so the spring crop was small, but then it rained in December and now this crop this year in Canterbury is the best I can remember, ever.

Steve: What, literally this year?

Peter: Yes. That's not being reflected right throughout New Zealand. Typically here we get winter rains which stock up the water table and we might get a few showers to keep things going in summer, but every so often we get a sub-tropical low come though, maybe every three years, and that can give rain at the right time.

The country had a really bad crop in the 1990s of about 4,500 tons, and that was because of cold conditions. The rest of the country does well if it's fine and dry but Canterbury does well if it's not, because we are susceptible to drought. If you look at Europe then hot dry weather creates a fantastic honey crop.

Steve: We don't often get hot dry weather where we live! We live in Manchester.

Peter: Our latitude is pretty much the equivalent of Spain in the Northern hemisphere.

Steve: That's nice. It's where we often go on holiday. I was told that the best weather for a good nectar flow was mist in the morning and then it warming up later on.

Peter: It can be but clover, which is our main crop in Canterbury, seems to do well with really hot and dry conditions during the flow. This year will be perfect for it. We have had a big growth in dairy farming here and that tends to be incompatible with clover. They like really thick lush grass and they use a stick to measure the depth of grass, and if it isn't thick enough they add water and fertiliser, and then the grass dominates. Clover's never going to grow on heavy soil but in places with light soils and stony soils we get clover. In the hot and dry conditions they can't keep up with the watering so the clover comes through and the bees get a lick. That needs really hot conditions; this year would be perfect for it. If it's a clover seed crop, where the primary crop is clover seeds, they will need pollination and so they try to maintain the soil so that conditions are right for a nectar flow, which will encourage bees to carry out pollination. Sometimes you get little short clover plants that yield phenomenal amounts whereas you can get the big plants that look fantastic but give you no honey whatsoever. It can look stunning but absolutely no production of honey comes off it.

We've got a site just down the road here where one of our local bee scientists was catching bees, painting them, then releasing them so that they go back to the hives where they would be counted. A local beekeeper found bees in his hive that had come from 4.5km away over clover paddocks. They had been caught in a particular paddock, painted green then released, and they were picked up 4.5km away. He opened up his hive and thought, "What's happened to my bees?!" The straight line between his hives and where they were painted had clover paddocks all the way along it. The bees kept flying over clover to get to clover, which can only be because the clover near the hives was not yielding nectar, so they flew over that to reach clover that was.

Mentors

Steve: How about mentors or people who inspired you, do you have any heroes?

Peter: I'd have to say Graham Kleinschmidt. He was my tutor at Queensland Agricultural College and he was a very far thinking and foresighted individual. He initiated research of his own, and I think he also inspired a character called Doug Somerville, who wrote a book called "Fat Bees, Skinny Bees."

Steve: Oh yes, I have that.

Peter: The contents of that book had their genesis very much in Graham's work on nutrition of bees and measuring crude protein levels in pollen. You've got the issue in Australia where most of the honey plants didn't evolve with honey bees. They've evolved by providing pollen and nectar as rewards to native pollinators like moths, beetles and native bees, parrots and so on. For European honey bees to come along and firstly find the pollen attractive, and secondly find it nutritious with the right amino acids, was fundamental to their survival. Then there was the discovery that bees can mine their own bodies to feed protein to their young during a dearth. It's a law of diminishing returns because if they can't get good nutrition and feed, the subsequent generation you get into a downward spiral. He nominated a number, a point of absolute no return, where the bees were done for. As the protein in the bees drops so does their life expectancy, so at some point you just get a complete collapse of the population.

When I came back to New Zealand in 1980 from Australia I looked around and thought that we don't have a problem here, but subsequently we did, because we now have monocultures in this region. We also have bees working on high yielding native plants but then failing the following spring, and it's primarily because of the poor quality of the pollen they are getting. Nosema impacts on this too, because it is a disease of the gut, and it prevents the bee from absorbing the nutrition it needs. If you have poor nutrition and a poor digestive tract it's a double whammy. If you've got Nosema but you have really good pollen resources you can get them over that hump; through a combination of eating the good pollen and mining their own bodies they can get to the next generation, and get past it.

Steve: Do you have both types of Nosema here?

Peter: Yes.

Steve: Talking about nutrition, what you are really saying is that bees have got to have a varied diet, aren't you?

Peter: Well, some types of pollen are fantastic, like gorse. It's a winter source, and any bees we had on sites near gorse would come into spring absolutely booming. Broom is a similar plant with a similar quality of pollen. It doesn't have to be a varied diet so much as having a really good source of quality pollen. Willow is another good one. Guess what? All three are European plants.

Steve: We do well with willow back home.

Peter: Our main one here is Salix fragilis, we call it Crack Willow.

My working week

Steve: Oh, here's a good one: how many hours per week do you work?

Peter: Not very many!

Steve: Oh, you're lucky then, aren't you?

Peter: Well, a lot less that I used to. I'm sixty now.

Steve: Do you own it, the whole thing?

Peter: Yes.

Steve: You're doing OK then. Are you quite relaxed or are there things that keep you awake at night?

Peter: Not so much anymore. I came in 1980 and took over from my father, and at times I have had tough years and sleepless nights. It's been pretty stressful at times. I've done most things here, and I've written the software used in the processing plant. We have a label on the honey which has a QR code, which is linked to a map of New Zealand and the location of the honey source. We don't do it publicly to the level that we can do it because beekeepers get sensitive about their sites, particularly Manuka sites. They are having to pay rent of $100 per hive plus a share of profits or share of the crop, and they don't want anyone to know where those sites are. We keep the detail low so that you can't see the exact hive locations.

That level of detail comes from software that I developed, which often took me into the small hours, writing it and tweaking it and getting it right. I still work on it from time to time, but that is just slight tweaks every now and then. I built a lot of the engineering in the company; I built the packing plant. Often there was a need to get the plant up and running the following day so I'd be working until midnight in the workshop.

Steve: But nowadays it's a bit more relaxed?

Peter: Yes, I've got a good team of people, and that's key. We have a profitable business which allows us to employ good people.

Revenue streams and the world market

Steve: Are you able to say anything about the breakdown of your revenue, roughly?

Peter: About two thirds of our turnover goes back into buying honey. The rule of thumb in New Zealand is that if we pay the beekeeper $10 per kilo plus sales tax the honey will retail at $20 per kilo including sales tax (15%).

So, the retailer will get $17.40 per kilo. Retailers' margins (markup) here are around 25–40% when all their costs/charges are included.

For our part we supply freight from the producer and to the retailer, drums, packaging, analysis, storage (interest costs) and labour to package the product and marketing and brokerage costs.

It's interesting to compare honey margins to jam. The raw material costs of jam, which are 50% sugar plus fruit, come out at around $0.75 to $2.50 per kilo. It sells for 3 to 4 times this much, whereas for honey the retail price is about double the materials cost. Honey is a low margin industry in New Zealand.

Steve: I wonder why the margins are so low?

Peter: Well, I think that beekeepers tend to be more socialist in their thinking. If they are selling to someone who makes a huge profit and starts driving around in a flash car...they don't like that. Also, it's a bit like the owner/driver of a truck; he wants to see his name on the door. He's driving down the highway going broke, but he's happy because his name is on the door! For every one that falls over there's another one to pop up and take their place. Beekeepers like to see their product in the market, it's apparently gratifying to them to have their product on supermarket shelves. They like to be vertically integrated, so they have the hives and they do the packaging and sell it to the supermarket. Every little town in New Zealand has a beekeeper like that.

If the world price is $4 a kilo and his cost of production is, maybe $3, then if the retail price is $10 he doesn't care if he gets $9 or $8 or maybe even $5 or $4. We've got beekeepers locally who will sell packaged honey to the supermarket cheaper than they sell to us in bulk. That's crazy; why would they do that? It's all part of that persona, and not wanting a middle man to make a profit out of them. It's counter intuitive, but that underpins the market right throughout New Zealand.

Steve: Right

Peter: So if you can't do what we do at a margin that is wafer thin then you are not going to rise to any great extent. Manuka has changed that because now suddenly you've got this story and the marketing and honey analysis and verification, which is a big part of the story, and which the little guy cannot do. People are willing to pay $30 per jar in supermarkets for a manuka product. The little guy can't do that because, for a start, manuka is not growing right across the country. As the supermarkets have got stronger it's changed the way that they think; they reduce the range of honey products on the shelves and become less interested in the little local guy.

We've also, in recent history, had a cooperative. Lots of people joined the cooperative and they didn't want anybody to make money out of them; the cooperative would pay out all of its profits every year. After a while they had to hold some money back, so they'd say that they would pay out after five years, but when five years came they'd ask the beekeepers to turn it into capital rather than be paid. It kept growing and growing, limping on, and they cannibalised their sales and made no money; lots of cooperatives of different forms around the world seem to do that. Eventually Comvita bought them out and then suddenly a real household name New Zealand honey brand, which was Hollands Honey, just vanished overnight. It's gone. To be able to compete against that...

Steve: That's tough, yeah.

Peter: We've had to innovate, to create different products, to create and respond to complexity. As the price of honey has gone up the local government authorities have decided that all beekeepers need to be on a register, and they charge $300 a year, thank you. They implemented more site inspections, and risk management programs, which all cost money. It goes on and on, and while the beekeeper is doing well because of the high honey prices they can afford it, but it's all cost that has little benefit. Have we had less consumers die? No, we didn't have any die before. What about getting sick? No, same story, they weren't getting sick before.

Part of this is the New Zealand persona. We are an agricultural nation; we produce food and we try to sell it into the world market. We are one of the most isolated nations on earth; it's three weeks shipping to our nearest major market, other than Australia. We've got all of this agricultural produce in a world that's basically awash with food. You know food's in plentiful supply because it's cheap. We have to compete with that and compete in markets that really don't want us there, they can make things difficult with border controls and risk management schemes and all those sorts of bureaucratic things. We are constantly fighting against that.

New Zealand has a philosophy of being as pure as the driven snow. We have to be able to provide assurances and products that are beyond reproach. We get swept into all of that, so the Ministry of Primary Industries, the MPI, have overarching legislation for the beekeeping industry because it's an "animal product". Then you go to Asia and they say, "we want no animal products in your honey." [Laughs] To them an animal is a mammal!

Steve: Well, in the UK, with Brexit coming up we'll be finding out all about that sort of thing. There are bound to be some changes.

Peter: Ha – renegotiating all of those trade deals.

Steve: Hopefully it means easier trading with nations outside of Europe, but I don't know. We'd better not get into all of that.

Peter: New Zealand doesn't sell an awful lot of honey into Europe now. We used to sell about half of what we produced.

Varroa

Steve: OK, what about varroa then? It's here isn't it?

Peter: Yes, it arrived in 2000.

Steve: Is it more or less everywhere now?

Peter: Oh yes, absolutely.

Steve: So it's the same as us; you treat every year and learn to deal with it...

Peter: We've had 17–18 years of it-and a lot of random inspections and sampling of hive contents to look out for any residues of medications or treatments above certain standard levels. It's remarkable that there have been hardly any detections, something like 5 in 25 years. It's an enviable record.

Steve: Yeah, very good.

Concerns and challenges for beekeepers (and the world)

Steve: Other than the weather, what is the big worry for you? I know a lot of hobby beekeepers and non-beekeepers are worried about neonicotinoids and pesticides, and apparently the bee population is collapsing and in about ten years we won't have any fruit. Many of the bigger beekeepers that I speak to say that varroa is their biggest concern, rather than chemicals. Obviously, there's foul brood and so on...

Peter: Yeah, all of those things are a concern. In the US there is an annual survey called the Pacific Northwest Pollination Survey which comes out around the time of the October edition of the Honey Market News, so snippets in there over the years show you where the real issue is. Starvation and queenlessness are the biggest cause of hive deaths worldwide, followed by varroa and other things. Varroa is only a cause of hive death in about 5% of cases on average. Does it cause costs and heartache? Yes, it does. I remember looking at our 6 year average honey yields before varroa and 6 year average after varroa, and it jumped from 30KG per hive to 36KG per hive after varroa.

Steve: So it's been well managed.

Peter: There was the question, "Why was that? Why did production increase per hive after varroa?" There were lots of theories, and nothing proven in any paper, but the miticides that were being applied to kill varroa were also thought to be killing pollen mites, so nutrition was improved. Beekeepers were commenting that the bees, "just seemed to do better."

We also had quite a large endemic population of Apis Mellifera Mellifera (Amm) and varroa certainly put paid to that for a while, and breeders were able to produce purer Italian stock, which typically was easier to manage and produced more honey. Then there was the thought that varroa got rid of all the bad beekeepers, but that didn't square with good beekeepers saying that their colonies were doing better. It was a very common opinion shortly after varroa arrived. You don't hear it now, because many of the beekeepers who remember pre-varroa days are gone now.

We had the possibility of bad beekeepers leaving, less competition from feral bees (in some places that couldn't apply because there were no feral bees), the pollen mite argument (which could only apply in some areas where pollen and food were a problem), so maybe there is some unknown, unexplained mechanism responsible for improved production. Maybe it was because beekeepers had to be in their hives routinely and more often than before, and they became better beekeepers as a result. All of those things could have contributed.

Right now our national average is 32KG per hive, but we've had a huge influx of new beekeepers and a massive increase in hive numbers, which could have caused production per hive to drop. Hives consume about 95% of all carbohydrates that they collect, and only 5% turns into honey on average. If you double the hive numbers in an area you double the amount of resource needed to feed those hives; the 95%, so it can become like having too many sheep per acre. We are beginning to ask what is a sustainable number of hives in New Zealand.

We have a new manuka standard that has been introduced by the government. There are lots of arguments about it, but one thing is for sure; there will be less honey that can be sold as manuka, particularly for export, than there was previously. How much less, I don't know. It could be 30% less or 50% less. We currently have an oversupply of honey problem in New Zealand, but the beekeepers doing really well can afford to keep hold of it. We had a bad year for manuka last year and one company lost $38 million. It was a joint venture and both parties lost $19 million each and have parted company. They had something like 40,000 hives. You can imagine what would happen if the world price fell to $4 per KG and they can't sell it as manuka anymore... "Who's going to buy our honey? We're out of here."

Steve: OK. Is the whole Neonic thing a big deal here?

Peter: No...it is where you get mis-application, and I believe that worldwide that's the problem. It's the thing that people can't quite put their finger on. There's no robust scientific proof that bees are dying en-masse because of endemic levels of neonicotinoids in the environment. There's lots of people saying we should be doing this or that, or eliminating this or that, but they are actually replacing much nastier chemicals that were there before but we never looked. So all the organochlorines and organophosphates have been replaced by neonicotinoids. Perhaps they have made it easier to have greater areas of monoculture, which is far more injurious to bees; I mean, if you've got vast acreages of corn across half of North America – there's not a place for a beehive. That's far worse.

Neonicotinoids have been credited with a 15% increase in the productivity of corn. Another way of saying that is that you now need 15% less land to grow the same amount of corn. I think of myself as being reasonably environmentally minded, and I think that the biggest cause of extinctions worldwide is taking land and habitat away from natural activities. We have a population in the world that's about 7.5 billion now, and the most recent estimates are that it will grow to 10 billion, before it falls away again, which means we've got another 2.5 billion people to feed. We have to feed more people without taking more land. It can be done, but the more land we take from the environment to feed that extra 2.5 billion, the more species we're going to lose. Man's impact on the environment is absolutely huge.

In Australia and New Zealand we've got fragile environments, but if people stop doing stuff, in ten years you can't see where people were – the damage doesn't last, it's gone. If you have an extinction that's it, gone forever, you can't replace that. To be good stewards of the world we have to stop those extinctions or minimise them, and we have to do that by taking less land, and neonicotinoids and GMOs, in my opinion, are tools that we have to get us over that hump.

Steve: OK

Peter: A really good example is India. In the 1950s India had 400 million people and was subject to famines and deaths from starvation. It was predicted that India was going to basically die out because they couldn't feed themselves, then there was an agricultural revolution where they introduced short stemmed varieties of the grains, new fertilisers, and improved practices. Just as the forecasts of doom were hitting the press they had the biggest crop on record, and their biggest problem was where they were going to put it all. Now India has 1.2 billion people and they are a net exporter of food, and they are using less land than they were in the 1950s.

Types of bees

Steve: OK, so the main honeybees that you have over here are Italians; that's right isn't it?

Peter: Yeah. There's a few Carniolans and we have Apis mellifera mellifera, and if you get a cross between those and Italians they are really nasty [laughs]

Steve: That's pretty well what we deal with. They're not that nasty, but it's all relative isn't it?

Peter: Aw, there's places in the on the West Coast of the South Island where there's been a lot of black bees and we used to run hives over there. We'd take them over there, and in the hives that swarmed the virgin queens would mate with the local bees. We had one area where there were 360 hives in this valley and most of the hives that came back that had swarmed showed up with black looking stock that was just nasty. Those black drones out competed the Italian drones hand over fist, and we had swamped that valley, every year for 30 years, with stock that had swarmed off – why didn't they dominate the area and stop the black drones mating with our virgins? They fly at lower temperatures and darker conditions, and jumped the Italian virgins every time.

Steve: In the UK we have these different camps; the black bee people and the Buckfast or Carniolan people – they really get quite angry with each other about it. "How dare you bring your bees into my area?" and so forth. The thing is, they are going to mate on the wing and you can't easily control that, can you?

Peter: They certainly do, and they go phenomenal distances to mate. When considering varroa here, and the spread of it, I thought that they had their calculations wrong. They said that they were going to create zones of about 13km or 15kms wide which varroa hadn't crossed yet, but they didn't take account of how far swarms can fly. There is little research on how far honey bee swarms travel but there has been work on African bees, which are known to have travelled up to 200–300kms a year!

Steve: Wow

Peter: So you've got an animal that's the same size and physical capability going 200–300kms and honeybees which are known to go 10–15kms but with no real studies on it. The other one they didn't look at was drones. I remember a beekeeper in Denmark saying that they had marked some drones and released them and he said that just using an informal study amongst friends they discovered drones that had gone 25kms. A drone with varroa on it will drop into other hives along the way, and nobody really knows how far they will go. If you try to prevent the spread of varroa by throwing a ring around and preventing movement, and you've got an insect that will go at least 25kms – who knows, they might go 100kms – then you've got a flawed model for preventing varroa spread. They seriously considered culling a lot of colonies as part of the exercise but they hadn't considered how far drones or swarms can travel; it would have put huge numbers of beekeepers out of business and not achieved anything, but thankfully it didn't happen.

Working with beekeepers

Steve: How many beekeepers do you deal with?

Peter: Off and on around about half of them. We've got routine loyal suppliers that always sell to us, then occasional ones who do some years, and then we've got people that float in between. We have a national apiary register, so to be a beekeeper you have to register once you get to a certain scale, and as part of that you have to provide map references. When beekeepers sell to us they sign a release form allowing us access to their apiary register so we get map coordinates, so we know where all of our honey comes from. We have around half of the apiaries in New Zealand on our list.

Steve: I think there is a lot of exporting of queens, is that right?

Peter: There has been, particularly to North America, but it's getting less and less now. There are two companies that still do a fair amount of it, and package bees too.

Steve: They have a long way to travel from here don't they?

Peter: Yes but it's worthwhile for them. After our honey production we have a lot of surplus bees that are just going to die off, so if they can take them, give them a queen and send them to the other side of the world they can extend their usefulness.

Steve: Have you ever heard of Rae Butler?

Peter: Yep

Steve: What do you know about her?

Peter: She used to work for us.

Steve: Did you part on good terms?

Peter: Reasonably, yes. She worked for another beekeeper up the way.

Steve: She's working on breeding VSH bees at the moment.

Peter: I haven't kept touch with that sort of stuff but hopefully some good may come of it.

Preditctions for the future and for Airborne Honey

Steve: OK, how about this: how have things changed from when you started to now, and what predictions do you have for the future?

Peter: From a beekeeping perspective?

Steve: Or from an Airborne Honey perspective if you like, whatever really.

Peter: I see that agriculture is going to get more intensive. We have seen that with dairy. There is a lot of land here that used to be dry land for sheep farming with really low productivity move to incredible productivity with changing over to dairy farming. That led to consequential productivity for beekeeping. There is much more cropping and also things like small seed crops; things like carrot seed pollination, rape and mustard seeds and clover. We have Borago officinalis (Borage) and evening primrose oil and many crops like that; some come and go according to fashion, phacelia and so on – there will be an intensification of all that. Pollen problems will become a bigger issue than is the case now.

Steve: The bee nutrition thing you were talking about?

Peter: Yeah. There has been this huge willingness among beekeepers to pay for manuka sites but also they are now willing to pay for overwintering sites; sites that have got really good bee nutrition. They can see that if they've got ripping hives going into a manuka crop then it's better than having half strength hives.

From an Airborne perspective, for many years we have run a laboratory, because it's the right thing to do. If you are going to claim it's manuka or rata or clover, it actually has to be. For us it's a little bit like jam, in that if you like raspberry jam you know what it looks like and you know what it smells like and tastes like, and if you were given a jar of raspberry jam without a label on it you could probably recognise it. The jam manufacturer puts the words "raspberry jam" on the label and they put a picture of a raspberry on it, to further bring your senses together. If you spread that out on some buttered toast you know what it's going to taste like…

Steve: You're making me hungry

Peter: Yeah. If you then bit into it and it was something between apricot and blackberries you would be gutted. You would say, "I'm never going to buy that product again." Why should honey be any different? If you look at our labelling we put the flower on the jar, and we use a clear container so that you can see the colour, because different honeys have different colours. Rata honey is white, and if you've had rata honey you will know its distinctive flavour; I love it because I grew up with it as a kid.

Steve: What's it like?

Peter: You can't easily describe flavours. I can tell you that it's mild or strong, or has a tang or whatever, but until you taste it you don't know what it's like. There are no comparisons which make any sense at all. Once you've tasted it, and you enjoy it, you should be able to go and buy it again, and again and again and again. You should have your expectations met. If somebody claims a mono-floral product, that has a unique set of characteristics that you like, you should be able to buy it again. That's what we believe we do.

Steve: Does that require lots of blending by you then?

Peter: Absolutely! But it means that you can maintain good quality and consistency over time. You have to be able to measure.

Steve: That's your whole ethos isn't it?

Peter: If you look on our Trace Me facility on our website you will see that there are seven or eight parameters that we measure, including sugars and colour and pollen and so on, which gives us the ability to be able to categorically say what it is and what it isn't.

Steve: And it's going to be that every time?

Peter: Yes. For us the mono-florals are usually over 80% pollen from that one flower.

Steve: What is your favourite honey? Is it that rata stuff?

Peter: Yes. But I do like a lot. Rata would be my favourite though.

Steve: I don't even know what a rata is

Peter: There's a genus called Metrosideros, and Hawaii has one that's famous over there called the ʻōhiʻa lehua, which has a story associated with it about two young lovers – one was turned into the tree by the volcano god and the other into its flowers by the other gods, to keep the lovers together. Lots of places about the Pacific have it; there's one in California and we have several species here in New Zealand. We have a northern rata, a southern rata and we've got one called Pohutakawa which is considered to be our Christmas tree; it flowers bright red at Christmas time. They are all very similar in their honey flavours.

Steve: On the subject of the future, I spoke to Murray McGregor, and he was optimistic about the future. He was looking at his children taking over from him one day and he thinks that the government is becoming more bee friendly, and that legislation will come to improve outcomes for bee farmers; perhaps subsidies or some better support. He was quite positive about the chances of an improved legislative environment. It sounds like your government is doing a lot, at least in ensuring that the reputation is clean. Do you see a similar thing for NZ?

Peter: Not from a government subsidy angle. If anything, the government should pull back and leave the industry alone a bit more. Prior to World War 2 the industry had a levy in place to market honey, then when WW2 turned up they didn't need to market honey because the government took it, because it's food. Government departments never like relinquishing control, so after WW2 the moneys that were being taken from everybody including the packers was used to subsidise the purchase of honey by the controlling body. It was the government competing against private enterprise. In 1952 they decided that it didn't work so they created producer boards; we had an apple and pear board, and egg board and so on, and we got a honey marketing authority. That authority survived through until 1982. They were the beneficiary of the levy that was originally taken for marketing, but they never did any marketing!

All exports from NZ had to go through Kimpton Brothers in the UK, so the price was set by whatever Kimpton Brothers could get. If you look at any price charts of honey, world prices and NZ prices back then, NZ got the world price. As soon as the New Zealand Honey Marketing Authority lost control the NZ honey price started to climb above the world price, and it's done that ever since. The manuka phenomenon has absolutely ballistically taken off!

The start of all that was back in the late 1970s and early 80s when government controls on exports were being lifted. For example we do honeydew honey here – so we have beech forests, and we have two species of insects which are not dissimilar to Turkish honeydew insects, they are the same family, the Margarodidae insects, similar to what happens in the Black Forest in Germany.

I remember figures of 9 Deutschmarks per kilo from Germany when we were getting 2. The opportunity for specialising and selling niche products into niche markets changed everything. Suddenly you are not selling honey, you are selling NZ Southern Black Beech Honeydew, which has a range of characteristics which you can measure and elucidate, and make a claim about. Some of the oligosaccharides in honeydews have a pre-biotic effect – I don't know if you are following health issues and gut health?

Steve: I do take a little pill every so often...

Pater: So you are taking a pro-biotic.

Steve: Yes

Peter: A friend of mine is just starting to market a Swiss product, which is a particular strain of a type of gut microbe which has been clinically proven to reduce cholesterol as well as statins, which is a remarkable feat. There are other claims that can be made about many other of these bacteria, and they need to be fed with pre-biotics, which can be found in high levels in a number of different honeys. You have a known therapeutic effect from some types of honey, whereas a nutritionist would just say that it is "sugar" and therefore empty calories. That's bullshit; we do have known therapeutic benefits.

Things that rile me

Steve: OK. Is there any issue that you have a "bee in your bonnet" about, something that really needs sorting out in your opinion?

Peter: The manuka standard has been one that I've had a lot to do with over the years. I was involved in the early marketing committees. When the NZHMA relinquished control the industry said, "now what do we do?" We knew nothing about who buys honey, from where and at what prices. I was involved in efforts to quantify the marketplace; I was involved with the marketing committees within the industry that created the manuka phenomenon. I was closely associated with the PR person, Bill Floyd; I was his fact provider – he'd want to go off and make all sorts of outrageous claims, and I was saying, "you can't say that, it's not true!"

Early press releases suggested that manuka would cure stomach ulcers but that they were awaiting clinical trials. The industry then picked that up and took it forward. I see that while it has been hugely profitable for New Zealand it actually embarrasses me to say that we have done that. As soon as the evidence was no longer there it was clear that many of the claims were untrue. I can walk into any tourist shop in NZ and there will be someone sitting behind the counter, and you ask them, "why would I buy that manuka honey?" – the last time I did that they said it would stop me getting a cold; it makes you recover from the flu quicker. It's rubbish, just total and utter lies...

Steve: People want it to be true, don't they?

Peter: Yes, they want a silver bullet. They want alloe vera, or echinacea or whatever it might be, but today science has been corrupted by marketing, and I think that is a sad state. It embarrasses me that our country, which is trying to trade on it's international status as being absolutely bona fide cosher, has the manuka phenomenon as our dirty little secret. The process of creating a standard could have gone down the Codex line which would have been based on pollen analysis, conductivity and a few other things, but the industry didn't want it, because they wouldn't be able to sell what they're selling as manuka honey. A lot of the stuff that had "activity" wouldn't pass that scrutiny, so they didn't want it, and now "pollen analyses" in New Zealand is a term which is absolutely derided by these people.

They didn't know what was causing the activity, and they were told by researchers who went down the wrong rabbit hole that the higher the activity the better the honey. When they finally found out what caused the activity, and that the precurser was in the nectar, they could measure the nectar and see that some had lots and some had none, even though they were from the same species. How does it then follow that a higher level of activity says it's a higher proportion of manuka? If you take a high activity manuka and you age it, because they found that ageing it increased the activity still further, you could see that you could dilute it down to 10% of its original volume and it would still be highly active. It's got high activity but only 10% is from manuka. That explained why they were getting high activity levels in honey which showed no manuka using classical analysis methods.

They were thinking, "What do we do now? We are selling this active ingredient, and it's what people want, but it isn't even manuka." They decided to just keep on going down the road that they were on because it was propping up huge sales and elevated prices. The fact is that there is this plant, and this concept of "wholly or mainly", and it should be wholly or mainly made up of honey from that plant before you can call it by that plants name. If you then want to make additional claims it should be, "wholly or mainly, with bells on it!"

Steve: Yes

Peter: That seems such a logical position, but the industry has gone round and around and now, as of December, the Ministry of Primary Industries have released their final standard for manuka honey. It's based on 4 chemical markers plus DNA, and we don't know yet whether that's going to work or not, but it's problematic. No country in the world has done this. No country in the world has created a standard without the words "wholly or mainly" and we've done that. There is no assurance with the standard that it is wholly or mainly made from the manuka plant. The concept of "wholly or mainly" embraces the fact that in a hive you can't get 100% from one source, and nobody expects you to, but they do expect a predominance if that's is what you are naming it. That has been lost in our standard.

We've got huge opportunities for our other New Zealand honeys. We've got unique species here, found nowhere else in the world, and they make wonderful honeys. We can talk about how special and unique it is, and that there is a limited supply and you can't buy it from anywhere else in the world – those are unassailable opportunities that can't be changed. You can certainly take away the manuka mythology by saying it's actually bollocks and, guess what – it's carcenogenic and mutagenic. The DNA test they are using on manuka is failing for the higher quality honeys, and the reason is almost certainly because the methlyglyoxal cross links the DNA. It doesn't happen so much on fresh honey, but when it's been subjected to two or three years of ageing and growing higher levels of MGO, when they test that honey it has no manuka DNA in it. The level that they are measuring it to is so sensitive; if it has one pollen grain it it they should pick it up, yet it is not finding any. The MGO just chews up the DNA and cross links it, so you can't find the sequence that you are looking for.

Steve: Wow. That's quite complicated.

Peter: Oh, it's hugely complicated.

Legacy

Steve: What about legacy, then? Do you have family to pass things onto?

Peter: I've got family in the business at the moment, but I don't know. In many ways I'm not a traditionalist; I don't have any great feelings one way or the other about that. If they decide that's really what they want to do then they'll have to show that they can do it

Steve: Seems fair. What about your plans for the future?

Peter: What, you mean the business?

Steve: Anything. You do photography don't you. I emailed Kirsten Traynor of the American Bee Journal and she said that I must ask you about your photographs. I nicked one of yours for my blog; the clover field one, but I did put your name on it. Is that a big passion of yours?

Peter: I come and go. I went off to Africa with my partner last year and between us we took 7,000 photos.

Steve: You'll get one or two good ones out of that! [laughs]

Peter: There were one or two! I enjoy it.

Steve: So if you weren't working what would you do?

Peter: Oh, travelling, definitely. Photography definitely. It's a pain in the arse trying to transport all of my gear and my lenses around. I spend quite a bit of time in Australia. I've got a 4WD Land Cruiser and an off road camper trailer, so I've probably done 100,000km around Australia in the last ten years or so, photographing bird life and insect life, frogs and so on. It's the diversity; New Zealand has a real paucity of wildlife, we have some unique species but we've got 300 species of birds whereas Australia has 900. We've got 3 species of frog, and Australia has 240. The wildlife in Australia is much greater, and also Africa.

Steve: I guess that coming from here you would tend to gravitate more to the wilderness areas?

Peter: Yes.

Steve: We feel that almost anywhere we go there is more space than where we are from. It's always, "Wow, there's a lot of room!" People seem to be in less of a rush, it makes you wonder why people want to come and live in our country really. I always say to my kids that if I was their age I'd be off somewhere else, where you can actually afford to buy a house and start a business.

Peter: House prices have grown dramatically here too. It used to be 3–4 times the average income to buy a house but now it's 10+ times.

Steve: Oh, right. That's bad. So your holiday would be in Australia or Africa then?

Peter: Yes, but I'll probably get to Europe next year. My partner's daughter wants to visit 30 countries by the age of 30.

Steve: Wow

Inventions and adaptations

Steve: Some of the things which your company relies on to be successful were basically invented by you, I think it's fair to say? I normally ask beekeepers about any inventions they have made but it applies here – you have actually written code and made innovations.

Peter: We had to create a standard for taking representative samples of honey, because without a representative sample you might as well be analysing ditch water. We put these through our lab and decide whether or not we want to buy it on the strength of the analyses. We put it into our computer system and generate a bar code for that batch; it goes on every drum and creates a stock record. We can see all of the stock on our system, categorised however we like, and we can see from that which honeys to blend together to get a desired outcome. We create the finished product and put it back through the lab to compare the outcome with the prediction, to ensure that it is within acceptable limits. For example, we have done an analysis of about 100 manuka batches over several years and the average standard deviation was about 3.2% away from our prediction, which we felt was a pretty reasonable outcome.

When we show people what it is that we do and how we do it they walk away saying, "that's logical, everybody else must be doing that." We are in fact the only ones doing that and we created the process and systems to be able to do it. The software handles all of our requirements and encompasses all of the documentation, such as honey purchase agreements with suppliers, meeting all of the government standards and so on; without it we'd be dead in the water.

Pyrrolizidine alkaloids

Steve: Is there anything else that you want to talk about that I can include in my book?

Peter: There's an overall philosophy out there that the analysts are trying to stay ahead of the cheats, and there's a product called honey which isn't really honey at all – that's a big issue in the world market. There's an issue where the technology and analytical equipment doesn't do justice to everybody, and people get painted as wrong when they are not. In Moore's Law, where computing power doubles about every 2 years, that also applies to computers in analytical instruments so we now use tools that are much cheaper and faster than they were in the past, and we start to look for things that we never did before. One issue now is Pyrrolizidine alkaloids, I don't know if you have heard of them?

Steve: No, sorry.

Peter: Well Pyrrolizidine alkaloids effect your liver and you can get veno-occlusive disease which means your veins close up which can cause death. Apparently grains and seeds from some parts of the world like Tajikistan, Afghanistan and Ethiopa are sources of these chemicals, as well as some herbal teas from Pakistan and India, and there have been many deaths. There have been thousands of samples of honey tested and no deaths anywhere, but a problem is looming. The effects in the body are cumulative so you can't just look for it in honey; you've got to be looking for it in the bread, the cheese (because the cheese came from animals that ate the grains with PAs) and eggs and so on.

You need to look at the total dietary intake. The fact is that the human population isn't dropping dead from veno-occlusive disease, but next people are saying that these are genotoxic, so we are looking at increased rates of cancers. I think that the ability to analyse and see so much more in a product is both a blessing and a curse in the same breath. You see it with the neonicotinoids where they can measure at vanishingly small quantities, far below any biological effect level, and say "these are here, and they must be having some kind of effect." It's a hard argument to say, "no – come up with proof, show us." They feed this stuff to bees in a lab and show the results, but we can see that they died of starvation! How good at beekeeping are the laboratory scientists? It's what I said earlier about science often being corrupted by marketing, and some of the agendas that many people in the world have are just another form of marketing. Our industry is subject to that I think, and we don't know where the next threat is going to come from, given this analytical capability.

Some people have started finding Roundup, or glyphosphate, in honey. The arguments about glyphosphate are absolutely passionately held and are very hard to refute. I wonder what the next one will be, because people haven't looked; suddenly one will pop up, and that is a risk for the world honey industry. Equally the improved analytical technology will help with identifying adulterated honey (with sugar). Honey commands a price that's about 4 x sugar prices at the moment so there is big economic incentive to adulterate. The higher the honey price goes, the bigger the incentive, and many would argue that the huge increase of exports, particularly from China is partially fuelled by adulteration with sugar.

Steve: On a completely different level to that, I was having a haircut the other day and my barber, who comes from Algeria, was saying that they test honey by pouring in onto the sand; if it soaks into the sand it's not pure. Have you heard of anything like that?

Peter: Oh, I've heard all sorts of stories and theories.

Steve: I think you go a bit deeper than that!

Peter: Well we do. We run the sugars, and if you are looking well you will see things that are normal and things that are abnormal. Manuka fails the C4 sugar test. C4 sugars – does that mean anything to you?

Steve: Not really, I'm afraid. A million years ago I did a chemistry degree but…

Peter: Well, there is a test which compares the Carbon 13 level in the honey with the C13 levels in the protein (mostly bee enzymes but also some pollen and some nectar protein) in the honey, and manuka fails it because the MGO breaks down the proteins. It means that if you try to sell manuka to America, as we found out to our cost, it shows up that manuka has apparently been adulterated; it fails the C4 sugar test. This is a case where analysis shows up genuine manuka honey to be fraudulent, so you've got to be so careful. I remember reading a paper which looked at how samples of genuine honey taken from bee hives in Brazil were sent to Europe for testing, and something like 62% failed their (European) tests for genuine honey. Honey bees evolved in Europe with European plants, but as soon as you step outside of that and you get into plants in far away countries, like manuka, or in South America, plants which have no evolutionary history with Apis species at all, let alone Apis Mellifera, things go awry. Is it genuine honey or is it not? This thing called honey is such an amazingly diverse product; it's an incredible problem.

As our analytical techniques get greater and greater the temptation for people with limited knowledge of honey in its widest sense to apply their best science to it is going to be a massive problem. We've just experienced it with our manuka standard. I think that we are going to regret that and have some problems down the line.

Steve: The place that you do all of this amazing analytical work, is that on this site?

Peter: Oh yes, it's like a shoe box at the moment. Our fire took out a workshop and we are going to rebuild that and create a lab with much more space so that we can put more equipment and people in. We do things on a shoestring here because as I told you we spend two thirds of our turnover on honey and margins are waifer thin. We beg borrow and steal in order to get a proof of concept, so we don't go out and buy some half million dollar equipment, we buy a $10,000 dollar piece from the States and get somebody to fix it and use it. If it works well and we can develop a market using claims out of it we then, at that point, when we really know what we want, go out and buy the best that we can. That takes a long time; our development is slow but we are not over capitalising and we are highly competitive, and we are able to go where others can't.

Honey production

Steve: All this stuff is interesting, and it's about seeing that not everything to do with beekeeping involves sticking your hands in a hive, does it?

Peter: Ultimately beekeepers are profitable or not profitable, and the biggest influence for their profitability is what they get for their goods and services. If we do our job really well they will be profitable, and if we do it poorly they will not. If you take that chart I talked about earlier, where the NZ honey price bumped along with the rest of the world price because all of our exports went to Kimpton Bros, think of the difference now. As the NZ price has diverged upwards and away from the world price that has funded all of the capital expenditure and equipment in our NZ honey industry, and that divergence comes from the work that companies like ours do.

Steve: You've got some barrels here – have they got honey in them?

Peter: Those ones are full but we handle about 3,000 drums of honey a year, so that's about 1,000 tons of honey.

Steve: You were saying that it gets extracted somewhere else?

Peter: Yes, in beekeepers premises all over the country. They have facilities which have to be registered with the government and they run a risk management program and so forth, so its all highly regulated.

Steve: If some beekeeper in wherever, somewhere on the other side of the island, decide that they want to sell you their honey, you've got to have given them a code in your database. Do you just assign a new beekeeper the next number in the sequence and add them that way.

Peter: Yes, that's exactly right. That's our internal number. Most beekeepers will go through some form of centralised extraction process, most don't have their own extraction operation – once upon a time they would have, but not anymore. It's too expensive to set up and run because you can't do it on the cheap. I knew a beekeeper a long time ago who set up his extraction equipment for $1,500 and he could extract 100 boxes a day with it. You couldn't do that now; it wasn't in a food grade facility and it wasn't made with food grade stainless steel. He didn't poison anybody.

Steve: Have you heard of Mike Palmer in Vermont?

Peter: No

Steve: I went to meet him and I remember him growling about some faulty part of his extraction equipment which was causing him hell. I think over here he wouldn't be doing that would he? He'd be sending it somewhere else that specialises in honey processing.

Peter: Potentially. It depends on the size of the beekeeper.

Steve: He's about 700 colonies sort of size

Peter: Yeah, there's guys here of that size that will extract elsewhere, but many will have their own at that level.

Steve: I guess it just depends on what they charge and what makes most sense.

Peter: Yes. I think it must be about $250,000 now to set up a rudimentary shed and install the extracting line. Hives have shot up in price but you can get an awful lot of hives for $250,000, so you have to work out if the path to greater profitability is buying more hives or doing your own processing.

Prospects for young people wanting to become a beekeeper

Steve: For some young NZ guy or girl who wants to become a commercial beekeeper, right now it sounds like that might be pretty tough, would that be right?

Peter: To get into it? Absolutely. But it's going to be very cheap shortly, because I believe there's going to be a crash in price. We've got beekeepers saying they can't run hives if the honey price is anything less than $8 per kilo. Whether they have done the numbers properly or not, I dont know, but many of them are hugely over capitalised. They get into it and buy a new truck, so they get a brand new Isuzu 4WD truck, and the same apiary has a brand new V8 Land Cruiser and they've got new this and new that…

Steve: It's not going to last is it?

Peter: It's just the largesse that they have become accustomed to because of the manuka phenomenon. The beekeepers that used to survive back in 2000 were paid $2.72 per kilo on average for honey, so with inflation that would now be about $3.85 today. They were surviving on $3.85 then but I don't think there is a beekeeper in the country who could do that now; even the one's who were doing it back then have become accustomed to the new normality.

Steve: OK, thanks very much for that. Can I take some photos now?

Peter: Yes, come through and lets have a look at what we've been talking about.

Afterword on Manuka

More recently I asked Peter if there have been any changes to the Manuka standard. His reply was:

Peter: There is still no reference to wholly or mainly in the standard. There have been two iterations of the standard from its first release. Two chemical markers (of the four) denote some proportionality by having changing values between mono and multifloral manuka. In one of these the median value varied by 5.7 times in the nectar data in the two consecutive years samples were taken. In the other it varied by 13.7 times! MPI took no account of the variability of these chemical markers in the nectar when they came up with levels required to meet the manuka definition.

We are only into our second season for manuka under the MPI standard and it is thus very early days to say if this will work or not.

After the interview Peter gave us a guided tour of his site at Airborne Honey. At the time the production line was churning out countless jars of Manuka honey bound for Australia. I watched as the clear plastic jars were filled with honey, lids put on, labels attached and then the conveyor belt carried them to the packing room. They went into cardboard boxes stacked on wooden pallets; it was a smooth operation. I was delighted when Peter plucked a jar off the line and handed it to me as a souvenir of my visit. Later on I purchased some clover honey and some rata honey, both of which were delicious. In all honesty I'm not madly in love with the taste of Manuka honey, but it's good in tea.

I saw the technology that Airborne uses for warming honey to change it from a solid to liquid state. Honey barrels are put horizontally in cylindrical containers and spun at a predetermined rate and temperature; a process which has been proven to produce the lowest HMF levels across New Zealand. I also saw the laboratory and met the three staff who were working there. The main activities were analysing the types and quantities of sugars and pollens in the various honey samples received.

On the drive back to Christchurch from Leeston I was able to consider what else we were going to do now that the interview was finished. As we crossed the braided Waikirikiri River we decided that we would try the Lord of the Rings Edoras Tour up to Mount Sunday. What a place that was! I do not particularly care for the Lord of the Rings, although I do sympathise with Golem, but the scenery around the film site for Edorus was something special.

October 2019 update from Peter:
As a foot note, the NZ price of honey has finally fallen back to World levels with the significant oversupply of production vs available markets now unable to be sold as manuka due largely to the MPI manuka definition.

Prices have fallen to around NZ$3.00 per kilo. This is at the same time as the World market is also at a 15 year low, again for similar overall reasons but on a lesser scale. High prices induce beekeepers to increase hive numbers and feed more sugar (so they can remove more honey rather than leave it on for feed over the winter), both of which increase honey production. When supply gets significantly ahead of demand, prices fall. These are long term cycles of 10–15 years or more.

Another factor that affects this is adulteration – very much in the news at present, but not a new thing. When the price of honey is high compared to other sweeteners, there is a greater incentive to cheat, and increased supply from other sweeteners being added to honey also helps to saturate the market. Legal substitution of some honey with other sweeteners by manufacturers using honey as an ingredient also reduces honey sales and increases surpluses depressing the market. In countries like the USA where a large amount of honey is used in manufacturing, this can have a considerable effect.

Many beekeepers in NZ are now getting out of the industry and those that are left will have a tough time for some time to come. For the mean time manuka prices have held up reasonably well but I personally don't think the long term looks bright for manuka.

Lovely frame from brood nest

Richard with mating nucs

Richard Noel, Brittany, France

'What's fundamentally changed is the way I deal with my bees. I have learned how to go from worrying about them and faffing and messing about with them to actually managing them properly. That's how my work has changed. Yesterday we did a process of giving those mini nucs space and introducing nucs to hives. It wasn't, "Oh, let's go and look at the bees for an hour," which you do when you are an amateur.'

Introduction to Richard

I discovered that two of the beekeeping gurus that I had met in my quest for knowledge, Michael Palmer and Peter Little, had previously been visited by the "up and coming" Richard Noel. I was intrigued to understand how any imparted wisdom might manifest itself in Richards rapidly expanding bee farming operation in Brittany. His YouTube channel was becoming very popular with its no-nonsense, down to earth videos showing how he keeps bees throughout the year. Many hobby beekeepers think that they might want to go full time on the bees and drop their day jobs so meeting Richard was an excellent chance for me to get the views of somebody in the process of doing that very thing.

Another fascinating reason to pop over the English Channel for a few sunny days in early May was to find out more about the Asian Hornet (Vespa velutina). This bee predator has been established in France for several years and has only recently started appearing in isolated pockets in the UK. It is a big scare story in England; considerable resources have been mobilised to educate beekeepers about the threat and to eliminate any nests found. I wanted to ask Richard about how he copes with them and what he thinks about the possible risks to beekeepers back home.

My wife, Elaine, was not going to miss out on the chance to spend some time in rural France in late spring. We were delighted that Richard let us stay in his gite, which sits beside his beautiful stone farmhouse, adorned with wisteria and set amongst farms and fields. A large Hawthorne hedge divided Richards land from the expanse of oilseed rape next door; a vibrant carpet of yellow that was humming to the sound of bees. I don't know much about cows, but the field to the front was home to several of them; they were a sandy brown colour, and some had horns. An owl box was attached to one end of his house, and Richard later showed us video footage of the chicks inside being brought mice by Mother Owl.

I saw and heard several Asian Hornet queens up close as they went about their business. They are smaller than common hornets, and the impression I got as I watched them was that they are black. On closer inspection, I could see the yellow legs and a single yellow band on the abdomen. They make a loud low humming noise as they fly. I found them to be quite beautiful and non-threatenning, but I think it would be different if I disturbed a nest or if they were decimating my bee population. At this stage of the season, the queens are out and about setting up their first nests after emerging from winter hibernation. A big part of controlling their numbers in Brittany has been large scale trapping of queens in spring.

Richard Noel is a man bursting with an effusive enthusiasm for beekeeping. He talks a lot, and he talks fast! This, I think, is the right way to be for somebody growing into a commercial sized bee farming enterprise. The long days of hard work, sweating in a bee suit and enduring setbacks and stings, could not be endured without a love and passion for this way of life. As I interviewed him, sat at the table in his conservatory, Richard's daughter Molly was, for some of the time at least, studiously ignoring us and getting her homework done. Later on, I was delighted when she told me that the French word for Walrus is "Le Morse" because my blog is called The Walrus and the Honeybee.

Talking with Richard

On good and bad years and becoming commercial

Steve: Can you think of any amazing or terrible years that you have had as a beekeeper?

Richard: Well, in some respects this really doesn't affect me because I haven't put myself on the line fully. I have only just got bigger in the last three years, and made that jump, but I have always had my other job to fall back to. One thing I always say to anyone who's starting, from my point of view, you'd be an absolute fool unless you had huge back up to just jack your one job in and say, "Let's go!" Unless you have a family behind you with a history of it you are taking too much risk by jumping straight in.

I have grown my beekeeping business over the last three years to a point where now I'm close to putting myself on the line. I'm at a point when I can no longer do two jobs. I have had massive ups and downs but every year I've made some honey and every year I've made some nucs. Last year I made more nucs than I've ever made and this year I'm hoping to make more. Financially I'm not recouping much yet but I'm always investing in order to grow.

I can't really say yet that I've had amazing or terrible years because I've always had that stability. It's the key to remaining on the path. I think you've got to keep your options open and remain open minded. As you saw yesterday you do have losses but you just clean them up and move on. If you get depressed and negative about it you don't move on.

In Brittany it would be folly to just rely on honey; you also have to sell nucs, which are good money. It gives you that sustainability because you can either sell nucs or use them to replace losses. This year I had losses because it was a bad winter and I think varroa caught me out a bit, but I had those nucs to put into the hives I'd lost. I had always talked about that but this year I actually had to do it. I'm not through the woods yet but I have a nuc yard stocked full of brood factories so that I'll be able to make cell builders, and very quickly I'll have more nucs than I did last year and new queens.

Why I became a beekeeper

Steve: How did you get started?

Richard: I was at a point in my life where I had land and I don't really know why, but after seeing a guy who had bees I thought, "I'd love to have some bees." It was nothing commercial or that I was going to have hundreds of hives. I got myself two colonies at the bottom of my garden, and they were all perfectly painted and pretty. I look at those hives now and think, "My God, I had a lot of time!" [laughs]

So, I had two hives and then I bought another off the guy the following spring.

Steve: What was his name?

Richard: That was Charles Basset, and he was the local beekeeper. I went to his workshop. Because I'd been reading up on bees it was like going into Aladdins Cave. He was one of those wizened old beekeepers that had frames hanging from the ceiling and blow torches on the side...all these facets of beekeeping in a tiny shed. It was packed to the gunnels. I felt like Howard Carter probably felt when he opened up King Tutankamoun's chamber, you know? I walked away and thought, "I want to do this." It was like I'd been infected with something.

He also had this method of doing artificial swarms, where you move the mother hive away, put a new box in it's place, take two frames of brood out of the mother hive – it's one of the earliest posts on my you tube channel, the classic way they do it over here. Basically all the foragers go into the new hive and together with the nurse bees on those two frames they make a new queen. It works but you lose a lot of foragers and you do stall your main colony a lot. You do forfeit some honey but it works. I started making artificial swarms like that and within two years I had nine or ten colonies, and I didn't seem to get winter losses. Perhaps I had more time to look after them and I think the winters weren't too bad.

Then I met my colleague Christian, purely by chance. I was at a gardening client's house working for him and I saw Christian and wondered over to see what he was doing. There was this epic line of about 28 hives in a row; these old wooden Voirnot hives. I couldn't believe it. They are slightly squarer deeper frames and they reckon they produce better.

Steve: How far away was this?

Richard: This was at a clients house where I was doing my gardening, probably about 20km away. None of my gardening jobs are that far away, like my apiaries now, because you want to cut down on the travelling. Fuel is a cost. I put my bee suit on and met him but then I didn't see him for a year.

I had young kids and was busy and just didn't give the bees my full focus at that time. Then after I got divorced I just buried my head in beekeeping. I went to America to see Mike Palmer, I went to Peter Little the following year, I met Steve Browning and other beekeepers and started doing you tube videos, and it just exploded. I still had my gardening job but I had to be really specific about what I would do and when because I had so little time. I learned that when a swarm is hanging in a tree if you can't get it into a box within five minutes you are better off leaving it in the tree, or put up swarm traps – that works.

When you have swarm traps about 200 metres from your apiary if they do swarm they often go straight to the swarm trap anyway, because the scout bees will have already found it. It's a really good use of old equipment and you often do catch a few swarms each season. It's surprising how many swarms are colony issues; you don't think they have swarmed but they have. The way Charles Basset got his bees was by putting up loads of swarm traps all over the area. In 2016 I put up 40 swarm traps and caught 31 good swarms. When I say "good" I mean of a decent size, but when you take them back to an apiary and put them in a nuc box then you find out what you've caught. Many of them re-queened themselves because they had old queens and I got very little honey off them the following year.

How do you set up a swarm trap?

Steve: Just out of interest, how do you set up a swarm trap?

Richard: You could use any box and if you put brand new frames in there's a chance bees might go in it, but the best swarm trap is an old box. I use old nuc boxes, and I reduce the entrance at the front. I put at least two old combs inside. They are old but they are clean. Pollen attracts the wax moth so you don't want pollen on the combs. You can blow torch the inside of the nuc which helps with the smells. I use two dark old combs, one that's drawn but lighter, and two frames of foundation. It should be at about head height. Putting them up is easy but getting them down when they have a big heavy swarm in can

Preparing mini plus frames

Barrels of honey

Frame of honey

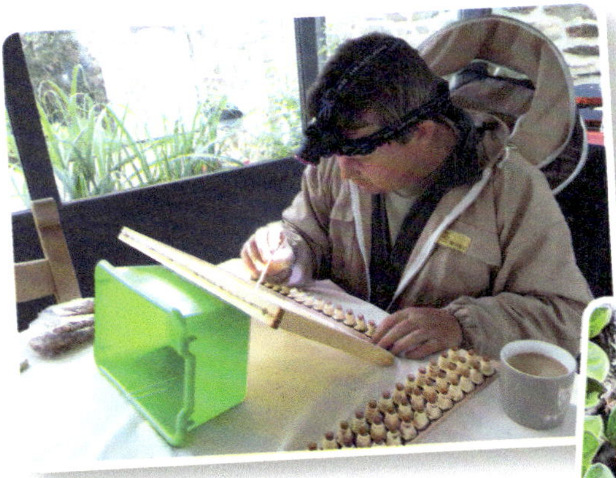

Grafting

Asian hornet queen

Making more nucleus colonies

Hive inspection

be difficult, so you have to be careful on a ladder. I close the box off and I put it on my shoulder, then edge slowly down the ladder but assume I'm going to fall. The centre of gravity is off to the side and the ground is unstable…it's a problem.

Collecting swarms in the summer is a nightmare. Bees swarm here between mid April and the end of July. You can't move them until the night, so in summer that can be ten o'clock at night. When you are swarm trapping you are out at ten and eleven o'clock at night moving boxes around, and because swarm traps work best without a ventilated base, you haven't got a lot of time. You close the box up, move it quickly within half an hour, then straight away open it up. It's a lot of work actually. Sometimes if you just put a swarm trap on a hive stand in the apiary it will catch one but bees are genetically programmed to travel when they swarm so you really want your traps 200 metres away.

If I get a swarm I let them build up then re-queen them in the late summer so that I get my genetics into that box, so they are not spewing out black bee drones next year to mate with my queens. It's not guaranteed but it helps; if you move bad bees out and keep the good stuff you eventually start to improve the genetics pool.

Anyway, back to how I do swarm traps: I also use lemongrass oil, but not a huge amount, and those little vials with the queen pheromone in. It is not a guarantee but it helps. Overall you need an old box and old comb.

Mentors

Steve: Have you been inspired by anyone or had mentors?

Richard: Obviously visiting Mike Palmer was hugely inspirational for me because I was also at a low ebb personally, and Mike managed to give me that lift, you know? I was just divorced and needed a break. I was able to bury myself in it. I was pretty sure I was going to be a professional beekeeper anyway but I could have just walked away from it. The day after leaving Vermont I thought, "This is what I want to do." He was just so cool. Mike's wisdom gets so ingrained. He's helped me but he's also said that some things you have to discover for yourself.

That was my biggest boost, but Charles Basset was obviously my main mentor. He had this little Citroen 2CV van; he was the classic stereotypical Frenchman, with the smoker lit so we couldn't see anything and bees flying around the van. I'm the same now, if there are bees in the van it doesn't bother me, but back then I was like, "Oh my God, there's a bee in the van!"

Steve: I'm still a bit like that myself!

Richard: With him it was like nothing. He just got older and I saw him less and less. I would have loved to get to know him better but I did know him well. I would go round to his place and he'd offer me a whiskey. In France they often have a drink at about 11 o'clock, a sort of pre-lunch drink. He used to teach beekeeping and run groups where people helped out making up nucs. Until you understand the fundamentals you can't progress, and I learned many of the fundamentals from Charles Basset.

When I went to Peter Little's he was doing similar things in a different way. The best way to become a good beekeeper is to visit as many people as you can, and ask them questions. Even if they are a bit whacky you can still understand what they are doing and why. They are not around forever, these people, and one day they are gone and all that information is gone with them. You have to seize the moment and suck up information when you can. You miss a lot; things change, you need to be on the ball all of the time. As you have experienced I can call up Peter Little and ask him a question, and I might come away three hours later with lots of different answers, then I can go away and think about it.

In some respects, doing my you tube videos has been my own mentor, because you look at what you are doing and question it all the time. I've never tried to be a teacher on the internet; all I do is talk about what I do and then discuss it with others.

Steve: With the you tube videos, presumably it takes a lot of filming to end up with just five minutes of the final product?

Richard: It does. For instance, when I did "The cell builder explained," that took ages, and I had rows with Molly [daughter] about what was in it. Molly knows how to raise queens now. She got me into the editing side of it and now I can do it on my own. It took a lot because I wanted to do a video that tells the whole story, but you can't because it's so complicated. I've had great feedback on it which is such a compliment. All queen rearing is basically a swarm box starter and a queen right finisher; they are all some kind of variation on that. I had somebody thank me because they made sixty queen cells, which is fantastic.

Steve: Then you've got to find something to do with them!

Richard: That's right, it's all about planning, and you need to do it and experience it yourself.

Steve: You've already more or less talked about how to get started commercially, in that you said to start slowly and not give up the day job.

Richard: Yes. I couldn't have had a better partner than Christian because he works so hard and has been at it for so long. He's made me completely aware of what's involved, and it's just work. It's moving boxes from one place to another, filling them up, emptying them out, picking them up again, cleaning them out and doing it all again. That's really what it is, but in a timely fashion with bees in them. You do need an awful lot of material.

Working hours and revenues

Steve: How many hours a week would you say?

Richard: In the winter, probably 20–40 hours a week, because I'm in the workshop if it's raining. In the summer I probably do about 60–70 hours per week and then I fit gardening in somewhere. But some weeks I'll be doing less, some more; it depends on the bees and where they are up to. I've got a guy called Ben Moore coming over from Australia to help me raise queens, make frames and so on. You never really stop beekeeping because you are thinking about it in the evenings too. Think about Peter Little, when he's not beekeeping he's on the forums.

Steve: It's an obsession

Richard: It is. It's like, "The more I know, the more I realise I don't know."

Steve: I don't know if this is too early to say, but what about the split in your revenue between different products?

Richard: I don't sell queens at the moment because I want to re-queen my existing colonies as well as making nucs. Mike Palmer, as you know, will re-queen in July because one of his fundamental points is that you can't expect an old queen to perform in its third year. You have to routinely re-queen. The key is that she is young and vigorous, that's more important than amazing genetics. I think it's best to have a young queen in the colony at the end of the summer because she will probably survive winter better and get going quicker in the spring. If you sell your queens for a short term gain you might be left in the spring with losses that you can't replace.

So, my revenue is not queens. Nucs and honey is about 50/50 at the moment. We don't have big honey crops here, you can't depend on it. With 250 hives in some years you can make 6–8 tons, which would be really good, but not always. Because our flows are so short and so start/stop, if you don't have the bees ready when it starts you can easily lose a super of honey. We have to feed them at certain times to make sure they are strong and ready for when the flows come.

If you don't have your bees fed and ready for when the flow starts you have to wait a week before they start, and that's a super of honey lost. If you look at it afterwards; you've got your bees here that you fed, which cost 4 euros per hive, and you have 2 supers at the end of it. The the bees here, which you didn't feed, you only have one super of honey but you kept your 4 euros. Which would you choose? It's a no brainer. At the time you think, "It's a lot of time and a lot of money," but you've got a long winter to come. Having that extra super on every hive is double your profit. The bees were ready to go, they had more brood, then when that hatched out those nurse bees helped more so the others could forage. Because you fed it in the middle of May they were stronger so when you make a split after the flow they are stronger and survive the winter better. It all compounds, you have to feed if they need it and you need to know your flows.

Steve: Do you use a brood box as a super or something smaller?

Richard: Our Dadant hives are a certain depth and the supers are half the depth of the brood box. That's how it works for us. In America with Langstroths they have shallows, mediums and deeps; we don't have that here, we have a super which is half the depth of a brood box. I run 10 frame Dadants and in the spring when we open up the hives they are on 7 frames.

The winter configuration is: a frame of foundation, a partition, then the brood nest on 7 frames, then another partition. Each partition is foam filled frame. When we make our last split in the summer we take 3 frames out, put 2 partitions in and a frame of foundation. They will pack those 7 frames with ivy honey which is enough for them over winter. In the spring when they start to grow they can go around that partition and start building out that foundation frame. Once we see them building that foundation out then we know they are off and running. We pull out the partitions and replace with foundation, and once they start working that you need to get a super on. If you don't they will start swarm preparations because the brood nest gets clogged up. They key reason for starting swarm preparations is a clogged brood nest. We wouldn't put the super on until those two new frames are being built.

Our winters aren't that cold generally, but even after this unusually long cold winter our production hives have still got honey in after the winter. We had quite big losses this year but it wasn't a feed issue, it was other things. In the spring what we do works really well. It's a good way of rotating your brood nest because you are putting new frames in. When we make splits in spring in a few weeks time we'll take 2 frames of brood out and replace with 2 frames of foundation. We are only making small splits in spring because we don't want to weaken hives too much and miss the main flow. We take 3 frames out in the summer.

When we take those three in the summer we then go down to 7 frames. All they are having to fill on the ivy flow in the autumn is those 7 frames, but sometimes if it's a strong flow they do build the frames on the other side of the partition. That's magic, because everything is big and packed, and when that happens the queen slows right down on her laying. We don't want them to have too much brood later in the year.

Correct supering is a really important thing. In America they often use a double brood box configuration which allows them to rotate boxes in the spring which gives the queen room and helps stop swarming. For us we seem to get away with having just the big bottom brood box. By doing a split in the spring you take away some bees and brood so at the critical time you do give the queen some more room. When we make that split in mid May the nectar flow is gone. Once you put in those two frames of foundation and feed them a bit she will fill them full of brood, but those bees that will come from that are not your foragers for the summer crop – they've already been born. If you're not doing a split after the spring flow then you really have to put a super back on to give them room; they might not do anything with it except sit in it, but it's room for them. At the moment I'm using all of the resource that I've got. I probably make slightly less honey than I could because I'm making lots of nucs and growing my numbers.

On finding apiaries

Steve: How many apiaries and hives do you have?

Richard: At the moment I have about 180 hives full, but I have another 80 hives to fill, so I'll end up with 260 full production hives at the end of the year. I've got 300 nucs but not all are full. I over wintered 120 nucs, I sold some, and I used some to replace losses and some have gone into new hives. By this year I wanted to be up to my goal of 250 to 260 hives which is what one person needs on their own to get an income. It's very difficult to manage more than that on your own.

Steve: Yeah

Richard: But when you have two people you need 450 to 500 hives to make an income, which is a big jump.

Steve: What about apiaries?

Richard: I've got about 14 apiaries at the moment.

Steve: How do you get them?

Richard: Right. I'm quite anal about my choice of apiary site. At first I was dead nervous about asking people, and I don't speak French that well. You have to build a rapport. The best way to do that was to ask someone if I could put a swarm trap in their garden. Over time I got to know these people and then one day I say, "You know that valley you've got? Does anyone keep bees near there?" Five or six of my apiaries have come that way. I have been on the site lots of times, and I've chosen them carefully, because I only put swarm traps up in areas that I think look good. I spend hours on Google Earth looking for areas that look good, and hours driving around in the winter. It's not just about forage; it's also important to have access and a good relationship with the land owner. I say to them that I will give them 4 KG in the spring and 4 KG in the summer, so it's 8 KG per year, which equates to about £100 worth of honey. If they were to rent out the sites that I use they would get much less in terms of monetary value than the honey I give them, so I think they get a good deal. Most people are grateful for the honey because it's fresh and local.

The thing to look for in a site is the access. If you can't get your vehicle in find another site because it's so important. You turn up and you can park near the hives. Sometimes you are too close, because sometimes they are grumpy in the summer when you do a harvest, and you can't get back in your truck because there's this great cloud of bees there [laughs]. Some of my places are on communal areas like old land-fill sites no longer in use. I had to write a letter and go to meetings but people are generally pleased to have the bees and the bio diversity.

I often agree on an apiary in the autumn but I'll do nothing and come back to them six months later and say, "Are we still OK to do that?" Having an apiary is a commitment and I cannot be moving bees when the flow is on. I've got my bees there because I want them to profit from that area, and if they say to me, "Oh, we've got a problem, we don't like it," I really really hate having to do the extra work in July. I cannot find anywhere else then because it's full. By that time I'm raising nucs. Often in a new apiary I just take nucs to it at first, then hives for them to transfer into. That way I don't have to move heavy hives. It also gives you time to build up the relationship with the owners.

If you have an apiary that's near a road and there are cyclists in the flight path then you are going to get grief. You have to really scrutinise it; is it cold in the winter? Is it overhung by trees? A little bit is fine but they need to get the sun on them, and not be dripped on by rain. Because I'm now involved in the Brittany professional beekeeper group the coordinator rang me up and said, "There's a strawberry farm nearby and the farmer wants bees – do you want to take a look?" I went down there and the guy could not have been more helpful. He showed me where he wanted the bees but I showed him another area that was better; no trees overhanging, sheltered, sun shines on them all day and the access was fine. Another thing; out of sight, out of mind. You have to follow all these criteria and eventually you find a good spot. The fussier you are the better the rewards, because everything is easier.

Steve: Do you ever run into problems with competition?

Richard: I have never had anything specific yet. I think that things like hives being stolen or pushed over is a result of direct competition, but I don't seem to have anything like that yet. Where I am we know everybody and communicate. There is no commercial competition here, because it's such a rural area. Last year there was one guy that I saw who had 8 hives near mine but it wasn't really that close; about 2 Km from my hives, and to be honest it didn't make any difference. He had black bees, which I wasn't happy about, but those aren't very productive anyway. I can't stop anybody coming in but I'd like to think that because I'm established here and Christian and I are quite thick on the ground then we should be ok.

If I put down a load of hives then find out I'm right next to someone else I say something like, "Sorry mate, I didn't realise – I'll take mine away at the end of the season, is that OK?" That's how you should behave. You should be gentlemanly, but unfortunately not everyone is like that. In time it always sorts itself out, because you have hardly any crop for a year and nor do they, then they realise that you aren't moving so they will never get a crop unless they move. In terms of amateurs the only thing I don't like is drones. We can't stop those black drones and most of them have black bees, which I might have sold to them so I can't complain! If the honey crop is good you can profit from that but if not you can always make splits and sell nucs.

On natural beekeeping and other controversial issues

Steve: There are some quite controversial topics that always seem to generate a lot of heat and I wondered what you thought about them? One of them is the idea of natural beekeeping, one is the debate about insulation and ventilation, another one is about importing queens and finally there are neonicotinoid pesticides.

Richard: Right, OK, natural beekeeping…I have fallen foul of voicing my opinion on it a few times, not that I feel guilty about that, but now I have changed my opinion. I don't think that there is anything wrong with natural beekeeping providing they don't have their hives near mine; my problem is when their mites go into my hives. There are not many natural beekeepers in France but there are feral colonies. They have done genetic tests on them and I think it shows that they are not the original pure black bees; they are a mish-mash of all different genes from different beekeepers and swarms. I would say that a natural beekeeper is someone who has a hive, puts it on the ground, and walks away. It's still not entirely natural because those bees would not have been there unless you put them in that hive. To me the best natural hive is a cavity in a tree, because they have room to grow if they want to, not like a hive which needs to be managed. The minute you put bees in a hive you have to give them space or they swarm.

You've got non-intervention beekeepers who just leave their hives alone. They swarm regularly and the colony remains small or it dies out through varroa. There's the other type with log hives hoisted up into trees, and then there's the sun hives and all that. The sun hive is actually quite small. All they really do is induce early swarming because they don't give the bees the space to grow. If you look at bees in a cavity, like a chimney, they can be enormous. What people don't understand is that a natural cavity for bees should be huge. The only thing I don't like about natural beekeeping is that if that colony gets a disease and dies then mine might get it.

We had an area once where I put some nucs there to be mated, near one of Christian's apiaries. His apiaries were clean but he got American Foul Brood then I got one case in one of my nucs. There was a place near there that our bees were picking up AFB from. We haven't had re-infection last year, or the year before, but the year before that we did. That was probably a wild nest that picked up the disease, or it might have been somebody's bees and they didn't look after them; there's no way to know. That's the kind of thing that can happen if you have a colony you don't inspect.

I consider bees livestock because I'm a bee farmer now. I feel that as soon as you bring in a hive you should check for brood diseases and varroa. If a colony dies from varroa, as that colony dies the mites know it's dying, so when your bees go in there to rob it the mites jump onto your bees, and they take them back to your colony. It's well documented in the UK that last year people treated for varroa and the numbers were right down, then suddenly they had loads. That's because in the autumn some hives fail and get robbed out.

I'm not in favour of regulation but I wish we could find a way of working with other beekeepers where we can say, "Look, I know you've got mites and it's causing me a major headache." Not many people here are "natural beekeepers," it's more in the South than here, perhaps because there are more old trees, I don't know. Fundamentally though, and this is important, I think everyone has a contribution to make no matter what kind of hives they keep.

We all need to share information, because it might be that someone who's doing natural beekeeping suddenly manages to breed bees that become resistant to varroa. Without that pool of people doing that I think we're missing a trick.

We have to leave our minds open and be open to possibilities. I don't believe that bees are going to evolve quickly against mites, because they are such a fundamental big problem. I think they will live with them, in certain pockets, but I don't think they'll ever be strong. It's like the small hive beetle; when you get them in colonies they constantly swarm, and they are small swarms and the bees are so agitated. People who have bees in a natural beekeeping environment have more swarms. The bees build up less, there are more mites, and they swarm more. I still think they are important to have because, as long as they are away from my bees, they provide an insight into how bees can cope. Maybe those bees have something that we will need in the future, you know?

What was the next question?

Steve: It was about insulation and ventilation

Richard: I won't talk a lot about this because for me we don't have winters really, we don't have deep snow on the ground. Mike Palmer, for instance, has three tiered hives to get the height, and they share their heat because it's one box next to another, and he has a ventilation hole on top so the bees can breathe. All hives in Canada that are outside have a higher vent hole so that they can fly if they need to. Over here we generally don't have much snow, not enough to suffocate the bees. I use the Dadant fully ventilated bases, that I keep open most of the year. If I get a cold spell and I get time I'll close the bases off in early February, to maybe help the early brooding by having less draft.

I think ventilation in a colony is a good thing but I certainly wouldn't have top ventilations here. The old idea of putting matchsticks on the corners of the crown board, which lots of people on the forums say belongs back with the dinosaurs, was to create an airflow. What you really want to do is make a pocket of air that stays warm, and the minute you make a hole in the top you get a chimney effect if the base is open too. In our hives in Europe where it's fairly temperate, we have some snow but not much, you are better off having a base ventilation. I believe that a solid base isn't as beneficial as a ventilated base, but that's just what I've found in my hives. I have used solid bases and find they get dirtier in the winter but there is a huge debate on it.

Steve: The thing is people who are making their living from this must have found what works for them…

Richard: Yes, it's area dependant. A lot of professionals I know all have the plastic Dadant bases with ventilation in. It's kind of a standard thing in France. I think it's great because you can sterilise them easily and you don't have to paint them. When something rots in a hive it's usually the base that goes first, and you get rid of that with one fell swoop by buying plastic, which is also cheaper. If you want to close them off you can; I need to do that when treating with oxalic acid.

On importing queens

Steve: Some people get hot under the collar about importing queens, but I think you've said already that you do that to improve the genetics in the area?

Richard: Yeah. The way I look at it is that I can't really work with what I've got here locally but I can raise good queens here locally. We have a mating station where we surround our area with our best stock, so those drones from previous generations mate with our latest generation queens. We are always trying to increase the pool and if we don't constantly bring queens in, and just rely on our own stock, there are so many black genetics in the environment that it just keeps going downhill. We have to keep bringing it in. I have to bring a good breeder queen or two in every year because if I don't keep making F1s the second or third cross gets nasty or less productive. Where Mike is, for instance, he has no wild colonies and he dominates his area with his bees.

If I've got a really good colony I'll select that queen for breeding so I don't just rely on imports. If I could do instrumental insemination here it might work well but I can't do that right now. You really need two of you to have the time to do it all, it's just non stop.

I don't disagree with these people who say that you shouldn't buy in queens but when you are in an environment where you can't work with what you've got without things deteriorating then that's the time to buy in a good breeder queen. I can see the arguments. There are people setting up a black bee project in the UK, which is interesting. If you've got people bringing in Buckfasts or Carniolans nearby that creates problems. For a commercial guy who just concentrates on having good queens that over winter better and are stronger and more prolific, you need to have good genetics. If you just relied on pot luck you would have less expenses but you would have half the honey crop and more disease and everything else. Don't forget, when you are buying in queens you are buying in hygienic traits as well.

Steve: To buy in breeder queens is quite expensive isn't it?

Richard: They are about 130 euros, but they are fully certified. We get an island mated queen which has all the paperwork. When that queen is producing all of those nice daughter queens it really pays off; the bees are prolific and calm and strong in the spring. People want to pay good money for your nucs and you get a good reputation. Everything becomes easier, the difference is so clear. I was really for the local bees but I got so annoyed with constantly getting badly stung and having to work in a different way. Things are so much easier now with good bees. You saw that yesterday. Imagine if I had to wear gloves, it's stinking hot, every time you get in the truck there's bees following you; we had none of that. It's only taken three years to get bees like that.

Steve: Yeah, they are nice.

Richard: They were generally good. There were a couple that got a bit more runny but considering what we did with them they weren't at all aggressive. Anyone can achieve that with straightforward culling of the bad stuff and bringing in good genetics. That's how I feel about it.

Steve: The other one was neonics?

Richard: France has just implemented a ban on most neonics here now.

Steve: The EU ban was on three types

Richard: Yeah, but there's now a new one that deals with more of them as far as I know – maybe you need to clarify that. The way I look at it is there's the least amount of insects and birds around at the moment than there's ever been in Europe. I said to you yesterday, we found some aphids on some ivy in the shed, which I thought was really weird because I haven't seen that for years. They say that the numbers of insects and wild bees and wasps are at an all time low. Something is causing that. I'm completely open minded about it.

I talked to Mike Palmer about this. They use neonics all over his area and he is a non migratory beekeeper, so his hives are right in the middle of places treated with it. He doesn't move bees, which takes away some of the stress and the single diet problem that you get doing pollination. His bees have a more varied diet by staying where they are so they can probably cope with the neonics better, if there is a problem with them. He says it's not just one thing; it's the varroa, the diet, the stress and so on.

If you have apiaries where the bees are well fed in a diverse pollen area they do well. I'm sceptical about saying that neonics are the fundamental cause of our problems, but something is causing it. I agree with what they have done. It's better to cut one thing out completely and then they can say, in five years time, this has made no difference at all. Then you know once and for all. You need a few years of a complete ban so that you can say, "Have the bee populations improved, has the birdlife?" We need to do that to test it.

I follow Ian Steppler's blog. He's a farmer and a beekeeper, but he is a commercial farmer dealing with annual canola. He sows that for his bees. He's an advocate for wildlife and bees as well as having to produce a commercial crop. He has to use neonics. I think that with advances in science we can make things that are so smart and specific. It's not the same as spraying broad spectrum chemicals all over the place, we can be specific in our treatment, and I think people forget that. We have to eat, so on the one hand I'm worried about it, but on the other we are managing to get more food out of the same area of land. I can see both sides. Half of me says, "Let's just go organic and stop all of the chemicals," because in a few years time the natural levels of pests will level out. When they treated in the 1960s with DDT it was the wonder cure, but it killed everything, all the way up the food chain to the raptors.

If this neonic ban doesn't work we've got a problem, because we still need to find out what's causing the population declines. I think it could be "all of the above." It could be varroa, bad climate and poor diet. If you ate roast beef you'd probably love it, but after a week of it every day you'd be sick of it. You can't give animals a single diet unless it's what they naturally do anyway. Our bees need diverse foods; they are designed to have varied diets across an annual flower cycle.

The beekeeping year

Steve: Can you tell me about your beekeeping year, as in what you do and when?

Richard: November to March is our winter break. We work in the workshop and try to structure our time so that we get most things done. For instance, this last winter I painted another hundred hives, but I won't have to do that next winter so I'll concentrate on getting mini-plus boxes ready or something. The whole key to my beekeeping is winter preparation. If I didn't have a chance to stack stuff and organise it and get it ready, come the spring I'd be in trouble. The honey flow is so short that you need to be ready for it.

In March we start doing our first inspections. It's just a quick inspection and if nucs need feeding we'll do that; a sugar solution of fifty/fifty water and sugar. It has to be warm enough to give a liquid feed, and this year has been unusually cold. In a typical year we would hope to get in the hives about the first week in March, give a light feed, and then the goat willow starts.

Steve: What is the temperature in early March?

Richard: 12 to 14°C. You might get one or two days that are 15 or 16. It's just a quick early look in the hives to see if they are alive and have a laying queen.

We don't do a full inspection; it's just lift up two frames and then leave it. We just need to know what's alive and what's dead, and assess their stores. We don't feed pollen sub because we usually have plenty of pollen in the hives from the previous autumn. When they do start flying there is such a huge amount of pollen coming in, it's incredible. They just pile it in. This year was completely different because they couldn't fly, but they still raised brood and look at them now. They are late but they have done well.

Around the mid to end of March we take out our two partitions as the colonies grow, and we continue to feed nucs if they need it. Going into April we start selling the nucs. We also start taking frames of brood out of the nucs to make up other nucs, if we've got the queens here. Then we are supering in mid April to hopefully get some honey crop. This year the blackthorn and hawthorne and oilseed rape came all at once whereas normally there is a slower transition from one thing to another. Sometimes by the time the bees are really booming the flow has nearly finished. It just depends on how good your spring is.

We are now in the first week of May so the flow will soon be finished. Then we start making our first splits and harvest what we've got. That goes on for the rest of May. By the time we get into June the colonies have had splits made off them, we've made nucs, and we are raising our first queens to go in those nucs. We are also feeding the hives that we have taken splits from and assessing what we need to do to get ready for the summer flow, because that one is key. The more colonies you have ready the better. The summer flow can be about a month long but often it's about three weeks. Last year was ten days.

Steve: Is that bramble?

Richard: Yeah, we have bramble which flowers mid June to mid July. From the second week of June it just goes mad. In a good year the chestnut trees will literally pour nectar out. The more colonies you have nice and strong by the beginning of June the more you will profit from that flow. You have to get the hives nice and strong, deal with the ones that swarmed and combine bees if necessary (a swarmed one with one that didn't). You have to be ruthless and just do it. The more checks you do at that time of the year the more chance you have of getting honey.

By mid July we have a short break because the flow is over and there isn't much to do. Then by the end of July and into August we harvest our honey crop. All through August we are making nucs, frantically making as many as we can, chucking bees into boxes and giving them queens. That's when you profit, because the bees have nothing to do then. They spend all their time working like crazy to collect nothing, unless the farmers have sown phacelia or

Stored honey supers

other late crops. So, we are making nucs and queens, we are re-queening colonies and treating for mites. It's usually amitraz strips but I've always used oxalic acid too.

You get that cross over point in autumn when the queen lays less but the mites increase, so you have to deal with the mites early. Get in there straight after harvesting the honey and treat them. In the eight week period between mid July and mid September there is nothing much for bees. It goes almost like desert here, really dry and arid. Some years are fine but not always. If you get rain in August the amount of ivy in autumn is incredible; the bees go mad and build out everywhere.

From the minute you harvest the honey all you are doing is getting ready for winter. You treat the mites in July or August then those queens start getting ready to make winter bees. Having those bees clean at that time is the key to having good over wintering colonies; they get away better the following spring. The ivy finishes around mid October to early November, and after that the bees will hardly fly. When it's mild they will fly out and get some hazel pollen in January or February, and some gorse; we do have a lot of plant diversity here which is great. I think the number one thing to keeping good colonies is having that diversity of pollen.

So that gives you some idea of our year. Some people have a longer season and some shorter. In California they are feeding pollen subs in January; we couldn't even begin to think of doing that. It's not really cold here but there are periods when you get an Easterly wind off Europe and it gets frozen.

Steve: What about keeping records?

Richard: It's something that I'm having to come to terms with rapidly. Now I'm a professional I need to be able to see what queens I've been crossing with. In France we are supposed to keep what they call a dossier d'elevage, which is a book recording what you've done in every apiary; when you visited, what you did, what you harvested and so on. I've got that for last year but I haven't written it up yet! It's a mess. It's so difficult as a beekeeper to keep records because you are so busy for those few months. I'm on my own, I have to be a father to my kids, a beekeeper to my bees and a gardener to my clients, and manage paperwork on top. I have an app on my phone which is helpful; it tells you where you've been in the day and how long you've been there. I go, "Did I really spend that long at that apiary!" [laughs]

Steve: Do you write on crown boards or something?

Richard: Yes I do, I generally write on the roof. I write things like "swarmed" or "took one frame of brood" or "gave a frame of brood" – that kind of thing. I don't write notes inside the hive. When we are mating we have a system of pins to keep track of queens. That's as sophisticated as we can get at the moment. If you put tags on hives they blow off. The writing on the roof goes faint and can't be read the next year. You could write notes on the crown board but it's just another thing to do as you are racing through hives. Sometimes I record notes on my phone because even though you think you will remember things you never do.

Steve: You talked about disease and pest management, so that's done, unless you want to say more about that?

Richard: Well all I'd say is that I've talked about having good queens and I believe that's one of the fundamental ways to help with your pest problems. You need strong colonies, and to have strong colonies you need good queens. If you have prolific queens the colony is strong. The queen can almost outlay the varroa, you know? You might have a lot of varroa but you have an enormous colony. It's that crossover thing again. I don't know what every breeder queen is going to be like, but even if she's not the latest in terms of hygienic genetics she'll be prolific, and they'll gather more pollen and nectar. In terms of disease if you breed out chalk brood you get more hygienic bees. If they are dragging out chalk brood infected bees they are dragging out other stuff, and that's why I think, touch wood, I don't have brood disease.

Whenever I catch a black bee swarm, what do I see after it's established? Chalk brood. They are just prone to it, I don't know why. That's why I don't work with black bees, because of all the other things I've discussed but they also get chalk brood! I use prolific queens and keep my colonies strong. All of the scientists and specialists say that no matter what you do, keep your colonies strong. That's the best way to fight all these problems.

Steve: Which is why you have to make sure that you are making up nucs at the right time, I guess?

Richard: Exactly. If you make them up strongly the queen will lay well because she is well fed and all the rest of it; it's the timing that's critical.

I bought a new piece of kit that will keep my queen cells warm when moving them about. I think it's the little things that I need to concentrate on now to try to enhance what I'm doing already. I believe that what I'm doing is the right formula, I just need to tweak it and get better at it.

Swarm control

Steve: Umm...swarming...that's a big deal isn't it?

Richard: Are you going to ask me about my swarm control methods?

Steve: I am. People split it into swarm prevention and swarm control. You've already spoken about your swarm traps.

Richard: I would say my prevention is giving the bees space at the right time and keeping queens younger. That's my two main things. Sometimes I can't do either. Sometimes they don't swarm; they are just not swarmy bees. I think that some bees are and some aren't, and I think that varroa is a key instigator of swarming. A varroa infested colony tends to swarm when the colony is smaller. If you keep your bees healthy and you give them space and super them at the right time you go a long way to help stop swarming. I know a lot of amateur beekeepers who are always doing this Pagden stuff and this manipulation and that manipulation, and build these gargantuan contraptions to control where the queen is...there's no doubt in my mind that taking the queen away from the colony for a short period of time is the best way to stop her swarming. How you do that is specific to where you are.

I think people probably think I'm an idiot because I don't think swarm control is a big problem, because I just can't carry it out. All the people I've spoken to who are commercial and know what they are doing say that there is no way you can do individual swarm control methods. You carry out a global management plan that overall effects the way your bees behave; you give them space, you super at the right time, you keep your queens younger. That is 70–80% of the way to help stop swarming.

If I go to a colony and it's swarmed or it's about to I'll bring a nuc along with me. If I go to a patch of brood where it's recently emerged, and there's fresh eggs, then I know she hasn't gone. You find her and then you take her away and make a nuc.

Steve: That's what Peter Little does.

Richard: Yes. If you find the queen is still there you can take her away. If you find that she's gone and there are closed swarm cells then it's game over. I put a note on the roof: "swarmed" and cut out most of the cells leaving one or two – I usually like to leave two actually. I'll come back later to assess the new queen and if she's laying like billy-o I let her get on with it. I don't worry about re-queening her if she is prolific, because they will be strong for the winter; a good strong queen from my genetics will generally be good anyway. Even black strong queens are good; I want to see loads and loads of eggs in a good pattern.

To be honest I don't talk about swarm control much because people say, "How can you not…," but you can't. It's regular inspections, keeping on top of things and giving space at the right time. By knowing the cycles and the nectar flows you can stay ahead of the bees, hopefully. I am bothered by it and if all my bees swarm now then it's a disaster for me, so I really do want to keep my bees in my boxes. There's just a limit to what you can do. You have to super in the spring and take some spring honey to unclog that core brood nest – it's just part of keeping bees.

Steve: In my amateurish ways I've come to the conclusion that, for me at least, when I do artificial swarms I end up with lots of small colonies and no honey. What's the point of that? Hopefully now that I've spoken to all of these clever people I'll have a better idea.

Richard: Yeah. What I was worried about when I started was that my colonies were too strong! I didn't have the experience to know that if I put another super or two on to give them space that it would be fine. If you can maintain that colony strength then when the flow comes along it's just amazing how much honey they can make. You can have all the qualifications in the world but if you don't look at your bees regularly you are in trouble; it's the difference between thinking what they are doing and knowing it. If you've got two or three hives and you have the time you can mess about with a Pagden or whatever. You might stop the queen from going.

Steve: In your case there is no moving bees to forage is there?

Richard: At the moment, no. There's a possibility I might move to a heather crop if we get a wet spring. The problem is, everybody moves there and there are too many hives. I believe in static hives. If you work hard and don't move your bees you can get two good honey crops. In Brittany we don't get huge crops but with strong hives you can profit from what we do have.

Steve: What are your main flows?

Richard: The spring flow starts off with all the small shrubs in gardens like chamilias but the real meaty ones are blackthorn, hawthorn, oilseed rape and goat willow. This is the perfect weather now for a hawthorn flow because it's warm and has been wet. We haven't had such perfect conditions for ages. Then you get oaks which only flower for a week but bees are all over them.

The summer flow is bramble, stuff like dock, and chestnut trees. You could stick a hive anywhere in Brittany and end up with chestnut honey.

Selecting queens

Steve: I think you've covered most of it, but what sort of thing do you look for when selecting a home grown queen for breeding?

Richard: I look for calmness, prolificness, they've got to be the best strongest queens, and we want strong build up in spring. Calmness is probably one of the biggest things because I can't work with grumpy bees or runny bees. Resistance to disease is a big one too. Also we want bees that get a lot of pollen. When I'm trying to do cell builders I really struggle sometimes to find good frames of pollen because the bees have used the pollen as fast as it comes in. If a colony becomes queenless thats a great place to find pollen frames because the bees are still foraging but there's no larvae to feed. You can find a classic frame with honey at the top and a huge band of pollen underneath. That's what you need when you are making cells. I'll often leave a queenless colony so that I can use it for pollen frames. You can store those frames for a few weeks with no problems.

On how I rear queens

Steve: I think you've got a really comprehensive video on this, but nevertheless can you explain the exact method you use for queen rearing?

Richard: Like I said before I generally go for the swarm box starter, queen right finisher scenario. I can raise a lot of queens in a short space of time. I have incubators. I don't use any chemicals; it's all set up to let the bees do it for themselves. It's a completely natural process. I'm kind of tricking the bees a bit; I have manipulated them by making them hopelessly queenless so they have to go into emergency drive. They flood the cells with royal jelly once you introduce them because you have set it up for them. You put everything into place and they do it for you.

Steve: With your starter you have frames of sealed brood in a box, but once they have emerged and done a round of queen cells do you keep it going by adding more sealed brood?

Richard: If I was using the cell builder to build one set of cells and on day five I re-assemble the colony so the queen is back underneath a queen excluder and the cells are finished in the top box, then when you harvest the cells you shake out the bees in front of the hive and then re-stock it with new brood combs, and on it goes. That's Mikes process, and every four days he sets up a new cell builder so he's got a constant load of cells coming out.

I use one cell builder but the next day I take out those started cells and put them into finishers. I use my starter box differently. I put the next batch of grafts in and move them to a finisher the next day and so on. I get more cycles through quickly just to start them. I use it for five or six days, get three or four loads through, and then re-assemble it. Then I set up another one using a different hive. I can get 150 cells out in six days, then I get them into the incubator so I can stock my first batch of nucs from that. You look at them all in the incubator and think, "Yes, they're done!" They won't all hatch out and they won't all be mated but most should be OK. It's brilliant.

There are so many ways of doing things. Last year Mike lost four batches of cells to virgins flying in. I said to him that's twenty days of production lost, and he agreed. I asked why he doesn't take the started cells out earlier and put them into finishers, because he has loads of potential finishers there, but he doesn't want to do that. I can't understand why, but that's his way. To me it's a much better bet to move started cells quickly to a finisher because once they are above a queen excluder they are as good as done, and no virgin queen is going to fly in there. I don't want to be telling Mike how to raise queens! [laughs]

Steve: Then you put cells in mini nucs to get mated.

Richard: Mini nucs and now we use normal nucs too. To be honest, I've been thinking about this a lot recently, and those mini nucs are a lot of work. They facilitate queens for harvesting, whereas nucs are nucs. I am beginning to think that putting a virgin or a queen cell into a nuc is the easiest way to work. Christian doesn't like me bringing loads of nucs to the mating station because each one is three frames of bees and brood, which is a lot of foragers in the area. The mini nucs don't take so much forage from the area. His argument is that using mini nucs means less feeding is needed in the other hives. I've started making nucs here at my place as well so they'll get mated here. The local drones may not be the best but I can sell the black bees in the spring as nucs. People want dark bees; the amateurs want the bees that need less work, so I keep the really prolific ones for me and sell the others.

Steve: What about introducing queens?

Richard: Right, well I like to do a live introduction of a virgin. First of all I prepare a nuc, give it four days then go and cut out the queen cells, so there's no more larvae that can be made into cells. I do put in my own queen cells but I prefer doing the direct introduction of virgins. I have queens in roller cages, they hatched out in the morning and I have marked them so there's nothing more to do with them. I just pick up a frame from the centre, lay it on the side and smoke it a little bit, then open the cage and let the queen walk out. She's a bit dazzled with the light because she's been in my pocket but she's accepted pretty quickly, and I smoke her and the bees around her. I get about 90% acceptance but my colleague Christian tried it and got a terrible result; I don't know why. Maybe I've just got the knack of it. He believes that it's better to use cells that are about to hatch. Again, it's what works for you.

Steve: Are you not often in the situation where you have to introduce a mated queen?

Richard: I do as little of that as possible. I'm probably going to buy about 40 mated queens from Italy to re-queen my last remaining rubbish colonies at the end of the year. We also harvest the last queens from our mini nucs to re-queen colonies and make good any nucs that didn't mate. Some nucs that you make up in the summer don't get mated, maybe about 20%. I want to buy those queens to enhance my genetics but also to make sure that every single colony is going to go into winter with a strong queen. I'm hoping that next spring when I've got my numbers up I'll be dealing with a lot of strong colonies.

I don't do much introduction of mated queens because most of the time I've been always making nucs to grow my numbers, but once I get to a stable number of colonies that's right for me it will be about re-queening from my own queens. When I get there I might be able to do direct introduction of mated queens like Peter Little does. I've done that here, but you need to literally take a queen out of one box and put it straight into the other box. I think they are better off under a push in cage because it's safer. If she's laying and is put on emerging brood that queen will be accepted. I so look forward to when I'll have better queens all of the time and not just be trying to make up numbers.

The reason that I've grown so quickly is that I try not to sell many nucs in the spring. You can make two splits off a strong nuc and those splits will grow and give you a honey crop. I can make so many more bees from strong nucs, and then those splits grow and can be split, and so on. If I sell a nuc it's 130 euros but that's the end of it. I see so many people selling nucs and then they ask, "How can I make brood factories?" By not selling your nucs! When you start off you have to try to build your resources rather than selling them.

Steve: Do you have a general rule of thumb about numbers of colonies per apiary?

Richard: Yes. In general, around here, you can't really have more than twenty hives in an apiary. You want honey in your supers, but sometimes they put everything in the supers and nothing down below, so you have to feed them. The more colonies you have the less forage to go around so you have to feed them more.

Steve: Do you have to move any to a different site for the winter?

Richard: There are a couple of places that I've got at the moment where I found I had high winter losses. It's quite high up and gets cold and the bees don't seem to do as well there. You learn. I don't think I'll over winter there again. I have found that in Brittany once you get to October it doesn't matter where you move your bees, you can have as many as you want on a site. In that small period after the ivy flow has ended you can bring them all into a sheltered area and then move them out again in March. They are easier to monitor because they are all in one place, check the feed easier – it's a safer bet. I'm probably going to bring 100 nucs here for winter this year. I can treat them all for mites at once; two hours work, done. It's a good thing to do. If you look at those German skep beekeeper videos he used to have hundreds in the same area for winter, and why not? The only reason you get disease spreading is if they are robbing but after the ivy flow it gets cold and they don't rob.

Steve: You said that your first inspection in March isn't really an inspection, it's a quick check of a couple of frames. Obviously once you are into weekly inspections do you have a time budget per colony?

Richard: If you are doing a full inspection it would take about three minutes per hive. That's taking the roof off, taking off the cover, pulling out the first frame, and working through the frames. The problem is when you have queen cells; that's what takes the time. Then you have to assess whether that queen has flown or not or whether she's there. We work it so that you can do five or six apiaries in a day. It's hard graft, but within a few days you've inspected all your hives. The spring harvest takes ages because we have to brush bees off the supers. We don't use clearer boards in the spring but we do in the summer.

Steve: I guess the reason that you can do a hive in three minutes whereas a beginner might take fifteen minutes is that you know what you are looking for?

Richard: Yes. First of all I see that it has a lot of bees, and I see if they still have space. Because I know what I should be looking for at that time of the year it doesn't take long. Straight away you can tell if they are queenless because if they are they are running around all over the place. As long as I've looked at two frames from the brood nest and I have seen nothing sinister and no queen cells, that's it. You cannot waste time. I see no point pulling that nest apart because if you've seen brood that's capped and you've seen larvae, and no queen cells, there's no reason to expect queen cells anywhere else in the colony. If you see a queen cell on the first frame then you think, "This is going to be a long haul." You've got to check every frame, find the queen if she's there and shake the bees off to find the cells. It really does add fifteen minutes to the time. You might need to make splits so you need to bring along extra kit and so on.

What have I learned?

Steve: From when you started to now are there any particular things that you have learned or methods that you have changed?

Richard: It's hugely different now. I love it now as much as I ever did because I feel more confident and I'm getting more results from my bees. I feel that I'm looking after them better and I'm able to exploit them better and more efficiently. When I started out I felt that I was muddling along. Now I've learned so much from other people that I'm better and I'm constantly trying to improve. It's changed now that I'm commercial, although I'm not really truly commercial; I'm nearly there but I really need to do three or four more years to get to be purely bees. Everyone who's commercial just does what I do, but they do it better and with more bees and more boxes, and more honey and more nucs and so on. It will come. You have to grow slowly like we said before.

What's fundamentally changed is the way I deal with my bees. I have learned how to go from worrying about them and faffing and messing about with them to actually managing them properly. That's how my work has changed. Yesterday we did a process of giving those mini nucs space and introducing nucs to hives. It wasn't, "Oh, let's go and look at the bees for an hour," which you do when you are an amateur. It becomes a different ball game but I never find it any less enjoyable; I absolutely love it. I just feel that when you are keeping bees you can live the life. You are doing a job but it's the best job in the world. It's hard and there are huge pressures but because you have a passion for it and are prepared to make sacrifices for it I feel that it's worth it.

The future of beekeeping

Steve: Have you got any predictions for the way things might go in the world of beekeeping?

Richard: It's an interesting question. Let me think. It feels to me that so many people are moving towards being environmentally minded, less reliant on chemicals, and that beekeepers will do well because they have such a natural product and such a good product. Basically we produce bees and honey and queens, and I believe that we're going to grow, and become more valued in our society. A lot of beekeepers plant trees and they know about pollination and are an important part of the countryside. What we do to help the environment directly effects the wellbeing of our bees so we are natural guardians.

A lot of people still see beekeepers as people who treat bees badly to exploit them, because there have been videos showing people moving boxes and splitting nucs with machines. That doesn't reflect what the massive majority of beekeepers are like and I find it really annoying. You get it in all walks of life. You get a plumber who rips someone off and you think that all plumbers are like that. The vast majority of plumbers are smashing people who work hard and fit bloody good heating systems, you know what I mean? I think that we do an awful lot of good and I'm so lucky in my life that I've had a chance to meet a few of the good beekeepers, and I hope to meet more. They have worked hard to develop in their area and help bees.

People are interested in what we do. They always ask if you get stung, to which I say "Yes," and they ask if it still hurts. I say it hurts like hell! They are all interested and ask lots of questions. Supermarkets are stocking more bio products and I think it's a good thing.

Steve: That reminds me: when you sell honey is it bulk or what?

Richard: It's getting the balance. This was explained to me when I first started. You can make a lot more revenue selling honey in jars with your own label on them, but the time it takes to get it marketed and looking pretty, and all the food safety standards and all that; is it really worth it? Is it not better to concentrate on raising good queens and making lots of honey and selling it in bulk? The majority of our bulk honey goes to other beekeepers who sell it on. I wouldn't want to sell it to a company that just uses it for salad dressings; I don't think that's good. I think my product is so good that I want people to enjoy it as honey. If I was part of a beekeeping family, like Peter Little for instance, you could have one person dedicated to the marketing and PR side of things. You can't spend hours and hours going to markets to sell your honey. It's a romantic idea but you won't make any money. You can sell some in pots to people that you know, of course, but for me right now it's mainly bulk honey.

Heroes of beekeeping and things that irritate me

Steve: Have you get any heroes from the past or present?

Richard: You can go on about Brother Adam but I never met the guy and I've heard mixed stories about him and his bees, but what he did overall made a massive contribution to beekeeping. It's not just about the Buckfast bee but the breeding methodology. People like him did a huge amount and it's a shame that he hasn't been recognised perhaps as much as he should have been.

Obviously Michael Palmer, because he took me under his wing and gave me a massive amount of help when I needed it. He showed me the road. I think Randy Oliver is a terrific guy; I don't know him but I have spoken with him on email. He has been massively influential in mite treatment and he will continue to do so because he has a big following and is a really clever guy who thinks like a bee. He uses science to make things better. There's other great beekeepers but I don't feel qualified to talk about them. There's AI Root, Dadant, but for me, it's great to read about the greats of the past, but really all you need to do is go back five to ten years to get the best and most up to date information. You always benefit from the wisdom of other people no matter how old they are. I really respect all of the information Dave Cushman put on his website. People always refer to it and it's the gold standard really. There is so much information out there on the internet and you tube now.

Steve: Is there anything that really annoys you?

Richard: Yeah. Facebook forums. They are probably the worst place to go if you are a new beekeeper because you ask one question and get a hundred answers. What annoys me the most is people who never read the previous posts. The same question has been asked and answered correctly time and time again in the past. All they want to do is see their name next to that post, which really bugs me. They haven't even got the manners to read the original post and the first few answers. What I also see in forums is that when a commercial beekeeper answers a question nobody says anything after that, because that must be the right answer.

Steve: That's what Murray McGregor says about his posts!

Richard: It's because what he says is correct, and he's a wealth of information.

He's done it all. I've never met him but I'd love to; he's a wealth of knowledge. I read Murray's posts because he passes on good knowledge in a positive way. That's another thing about forums; people don't reply in a constructive way. I posted on the BBKA forum, and got some grief for it, which taught me a lesson. I just do my you tube stuff now for people who want to follow me and I don't put it in anyone else's face. I get really annoyed and sometimes think I should come off social media.

Steve: What about if you had a magic wand and had one wish to fix some beekeeping issue what would it be?

Richard: One wish; to get rid of varroa. It weakens colonies and makes them susceptible to other problems. We have to use chemicals in our hives to treat varroa and we would never have to do that if varroa was gone. I know it's the obvious thing but it's the big one.

Steve: What about favourite book?

Richard: I would say "Honey Farming" by Manley because it's directly related to my work. It was terrific because it's exactly what a bee farmer does. It's old now but it still reflects what we have to do. One thing I liked was in the Forward where he says that he wished he'd become a beekeeper earlier on, because people doubted that you could make money from beekeeping. I've got some interesting older books but they don't reflect modern methods. I quite like that one by David Cramp because it's commercially minded and practical.

Inventions and modifications

Steve: Have you ever invented or designed or modified some equipment or gadget to solve problems.

Richard: One thing I did, that is not exactly groundbreaking, is that I made some ekes that the feeders can fit into, because I wanted the feeders underneath the cover rather than on top. I think it keeps the syrup warmer and it's more accessible to the bees. I don't believe that there's anything really new and innovative in the last few years. I think that it's pretty well all been done and once you find something that works for you then you stick with it. Those flow hives are innovative but it takes away the beekeeper from the bees; it's clever but I don't think it's good for beekeeping. You see people who invent new hives but to me it's like trying to re-invent the wheel. It's good that people try but I don't think we can make things much better than what we've used for years and years.

There are little things, like having plastic sheets on top of hives instead of crown boards, Christian taught me that.

Steve: Why is that better?

Richard: First of all you have to buy the wooden crown board, which is something like 12 euros each. It's wood so it will absorb moisture. The plastic sheet in direct contact with the bees and we can put insulation above it. When you want to take that top off you can just peel it back. It's a lot easier than using a crown board and in the winter the crown board absorbs moisture and its another space for the bees to heat. I think using a plastic sheet is brilliant, or a seed bag.

Steve: As we're in France, and I've already seen some of them about, and they have just begun to arrive in the UK, what's your take on the whole Asian hornet situation? I think they are very worried about it.

The Asian hornet

Richard: It's now year four of our Asian hornet "invasion". I did some videos on you tube at the start, talking about what I thought would happen, and it has come to be true. They are a problem and they have become established. As I mentioned to you before, our communes did a group trapping and they put out something like 5,000 traps across 60 communes, and I think we trapped something like 9,000 queens.

In 2016 we had a major problem. It was our first full year of them after they arrived the previous year, so that was the first full year of the queens making their nests after the previous year's invasion. We had a lot of hornets about. We didn't have losses but at the end of that year there was a lot of hornets bothering bees. We had to put screens up in front of hives and we were really worried, because it was all new to us.

Because of the trapping in the spring we then had no problem the following year. We all had traps in our apiaries and this group trapping by the communes was so successful that there were hardly any adult nests about, and the few that were there were found and dealt with.

Steve: So that's trapping the queens in springtime?

Richard: Trapping queens in the spring is the number one easiest solution. There's been a lot of research done, and if you allow a nest of hornets to reach maturity there are two issues. One is that there are a lot of hornets the following year. They are not just taking bees, they take other insects too.

Apparently one nest can take a ton of insects out of the bio-system – one nest! So, if you get rid of a nest by early trapping of the queen, even though you catch other things that aren't Asian hornets, you save many more insects in the long run. People are complaining about our traps catching common hornets, and we know we do catch some. We find that the Asian hornets start flying 3–4 weeks before the common hornet queens. We tend to get a much higher percentage of the founding queens in early March and April, whereas the common hornet is in April and May.

Early trapping is a very powerful tool, and these communes found that they saved a fortune because they had less nests to deal with later on. A large nest is a very awkward thing to deal with because there are so many adult workers defending it. If you disturb a common hornet nest you might get a few flying around and in the worst case scenario you might get stung. You've got to really poke it and piss them off to get stung. If you just disturb an Asian hornet nest they will chase you, they'll spit venom at you and they'll sting you. If you look at videos of pest controllers dealing with the nests it's a full-on attack, it's unbelievable, whereas with the common hornet they may not attack at all.

People say they are "killer hornets". One person who gets stung by a hornet is unlikely to die, but a person who happens to be allergic to bees and hornets, getting stung multiple times – they are going to be in big trouble. They are more likely to kill somebody because they are more aggressive, not because their venom is any worse. If you are allergic and get multiple stings you could die.

We've got them here and we know that if we trap queens in the spring they are less likely to be a major problem. Some people trap a worker and put a treatment on it, on its back. It's not a regulated treatment, it's a home treatment, but it gets rid of the nest, and you are just targeting that nest where the worker came from.

Steve: How do they apply it to a live hornet?

Richard: They use one of those electronic zappers, like those tennis racket type things, which stuns it. Then they pick it up with a leather glove on, paint the stuff on its back, and let it go. Believe it or not they are not that aggressive when they are away from their nest, because they are not expecting to be caught. It's a bit like bees. If you see a bee on a flower they aren't interested in you, they're away from their colony, but if you go to the colony you may get fireworks.

People say that you shouldn't do it because then you've got a dead nest in a tree that birds might eat, but you have still removed a nest from the environment, saving all those other insects, and the amount of chemical is so small that it naturally breaks down anyway. By treating a worker and letting it go back to the nest to poison the others they are specifically targeting the problem nest.

Steve: What is the chemical?

Richard: They can use different things but the one people tend to use is a fipronil based one.

Steve: They sometimes use that in New Zealand on bait to kill wasp nests, but you need a licence and training to be allowed to use it.

Richard: Regarding my thoughts on the introduction of Asian hornets into the UK, I personally don't think it will ever be a major problem. I don't feel that geographically it's far enough South to really raise big nests. The UK is a quarter of the size of France and has less unpopulated areas. The problem is when you get one nest that does reach maturity and it does release a lot of queens then, yes, a lot of queens will start the following year. But I think overall there will be so many people aware of this, that even if that does happen they'll do trapping the following spring to catch the queens. The process the hornet has to go through to achieve another nest somewhere else is likely to be halted somewhere along the line more so in the UK than here. We don't have the foot soldiers here. England has done a great job with the publicity and they have been proactive in dealing with the nest they found in Tetbury. They have done a lot of talks on it, and they have been able to learn from others in Europe who have had the Asian hornet for a while.

A lot of people in the UK are petrified that this is going to be a killer hornet but I can tell you, we are still keeping bees in France. The problem is that the further south you go the hotter it is, the more water there is, and Asian hornets love water. They hunt for insects on water. I think the queens have an easier time overwintering the further South you go. Quite a lot of them die here because it gets cold. In some parts of Spain it is really difficult now, they are inundated with Asian hornets.

The other thing I'd say is that whenever you get a pest arriving as an invasive species the first year it comes in it's incredible, it's so overwhelmingly strong, but then afterwards it seems to decrease. I wonder if the success of the trapping was also to do with the hornet being less prolific in its second year. I don't know. There's also the issue that these hornets are from a very narrow genetic line, and they haven't come from Asia as several queens; it's been one queen that founded a nest in the Girond and that's where all the others have come from. The scientists say that they reckon the hornets in Tetbury had the same genetics as our ones. I think that one thing that isn't helping them in Europe is that very narrow genetic line. But I'm not a specialist, and I'm only telling you what I see and what I know.

It's depressing seeing how many hives people are losing in Spain but up here we don't tend to lose hives. We have issues with hornets taking bees from outside hives but it's far worse further South. I think it's brilliant that in the UK they are proactive and going into it with their eyes open, and not panicking. I think there will be the odd nest but I don't think it will be a major problem because there are more foot soldiers and it will be well managed. I think there will be nests, and it will terrify some people, but after a while they will think, "Hang on a minute, we've had these for four years and nothing major has happened."

Steve: I saw an article in a bee magazine about beekeepers in France who put an extension onto the hive entrance, a kind of platform, and it had some kind of wire mesh above it...

Richard: Yeah, because they found out that hornets tend to hover a certain way when they are trying to predate bees, and if they give bees another route that the hornets don't like then it may help. What do I think about it?

Steve: Another gadget!?

Richard: Probably, but it's good that people are trying different methods.

Steve: Lastly can you say in a few words what the essence of beekeeping is to you? What's it all about?

Richard: It's all about being yourself and doing what's right. I can't be anyone else, I can only be me, and being me is beekeeping. I've always done what I think is right for me but I learn off others. It's so great to be able to do what I do, I feel blessed. I wish I had more time, like double the amount of hours in a day because I find it so frustrating when I can't do more with my bees. It will come and I know that if I stick to the formula and keep going in a straight line, and don't try anything too clever, and work hard, I will get there. I'm so lucky that in my life I have found something that I can do that I enjoy. There's not many people in the world who can go to work and say, "Wow, this is amazing." It's just brilliant.

Steve: OK, thank you.

Richard: Pleasure.

Randy Oliver, California, USA

'As soon as a queen producer can say, "I've got bees that don't need any mite treatments," everyone will be scrambling to catch up. I have set out a way for commercial queen breeders to do that. I'm a big picture guy. I'm looking at the whole thing and thinking, "What needs to be done? Not little band aids – how do we change the big picture for beekeepers worldwide?" I'm trying to fill in the holes to get that done.'

Introduction to Randy

On the morning of 24 October 2018, feeling out of sorts due to "jet-lag", I grabbed some coffee and muesli before checking out of my hotel. It was a place called Oxford Suites in the town of Chico in Northern California. I was only passing through so never got the chance to explore the place, but it seemed pleasant enough. It had that thing that you find across North America – the feeling that the motor car rules supreme, with drive-through restaurants and big roads and small sidewalks. Nothing surprising there. I did manage to visit their Safeway, which was enormous and full of wonders, and a diner called Red Lobster that served tasty shrimp tacos.

Just a few miles up the hill to the East, nestled in the forested foothills of the Sierra Nevada between Chico and Butte Valley, was a small town called Paradise. Little did I know that fifteen days later Paradise was to be completely wiped out in the so-called Camp Fire; the deadliest wildfire in the history of California. It claimed the lives of 86 people and the homes and businesses of so many more. What a shocking, horrific experience that must have been.

Of course, I was blissfully unaware of the disaster to come and was enjoying my drive to the Gold Rush town of Grass Valley, which is near to where Randy Oliver lives. The weather was beautiful, especially to a Brit in October, and I was getting the hang of driving in America. Luckily I had read up on "things to know about driving in the USA" because on arrival at Grass Valley I was confronted with intersections, or what we Brits call crossroads. In the absence of traffic signals at intersections the question of, "Who has right of way?" arises. I knew from my research that they adopt a very sweet system here; the car that arrives at the intersection first goes first, then the car that arrived next goes second, and so on. I'm pretty sure I annoyed a few people by queue jumping but I tried my best.

When I pulled up at my lodgings in Grass Valley, the Sierra Mountain Inn, I parked on the main road with two wheels up on the curb. A lady from behind the reception desk came out and gave me a withering stare. Her look of disgust was clearly directed at my parking position so I quickly moved the car off the pavement (sidewalk) before checking in to my little room with its own mini kitchen and en suite bathroom. It's amazing how different places have different customs and that small things like how you park mark you as an outsider.

The next day was again sunny with a glorious blue sky. I arrived at Randy's place, which is in a quiet and pretty area amongst hills and forests. A young fawn appeared, stared at me for a moment, and then bounced off into the trees. Randy Oliver is not somebody who can sit still for very long, but we started off chatting on his porch, enjoying a coffee and plucking a grape or two from the vines growing overhead on the pergola. The gentle breeze occasionally sent out melodic notes from the nearby wind chimes, adding to the rather zen experience. It wasn't long before he was up and about, and much of the following interview was made on the hoof.

Talking with Randy

Who is Randy Oliver?

Steve: In the unlikely event that there are beekeepers out there that haven't heard of you, would you mind introducing yourself?

Randy: Sure. I'm one of many hard-working beekeepers and researchers. I just happened to fall into a specific niche – the guy who's both a professional beekeeper as well as an informed bee biologist and researcher, who likes to share accurate information to the beekeeping community. I do this because I love learning and sharing what I've learned – it's certainly not for money or glory. My father was a well-loved high school teacher, and I've been sought out as a teacher in every trade and vocation that I've been in during my life. So, it was natural for me to start answering questions for beekeepers – based upon experience, evidence, and hard data, rather than dogma or "because it just makes sense." I get lots of letters of appreciation (along with donations) for being the "no-bullshit" guy.

To the great amazement of myself and my beloved wife Stephanie, beekeepers started looking at me as some sort of celebrity. I'm uncomfortable with celebrity – it's not something that I enjoy. What makes me feel most fulfilled is the appreciative feedback that I get from beekeepers and researchers worldwide.

A perfect example would be the beekeeper who I called to purchase two hundred pounds of pollen from for my current trial. When I asked for how much to write the check for, he said "Randy, we wouldn't still be in business if not for your writings – please accept it as a token of our appreciation." That's the sort of acknowledgement and appreciation that drives me. The other letter that sticks with me was from a nun in Estonia, who thanked me for being the main source of accurate information for the small beekeepers in her country

– the warm feeling that I got from that letter makes me smile – I'm so proud to be helping others worldwide.

Steve: Great, thanks for that. I was at Olivarez Honey Bees yesterday. They seem to be doing really well.

Randy: Three years ago he hired a consulting firm to teach him how to run a business very well, and he's got Tammy; Ray and Tammy are two of the most fantastic people on the planet, I just love them both.

Steve: I met Tammy very briefly and she seems really energetic and...

Randy: Yeah, and not only that, but the two of them so care about their employees. Their employees are like family to them. Every Friday afternoon they have a barbecue for all of the employees out there. They set up the grills in the kitchen and make a whole lot of really good Mexican food – it's just a good time. What Ray and Tammy like to do is not just a business, they are creating a supportive community of employees. When I see people who are successful in business who do it without stepping on other people, I really respect that, and that's Ray and Tammy. They are just the best.

A family business

Steve: I enjoyed meeting them and their children.

Randy: Yeah. That's the other thing, the children. What we are having is a whole generation, my generation, the pre-varroa beekeepers – we are getting to that age. When our kids get to their twenties they all say, "The last thing in the world I want to be is a beekeeper," and they take off and do other things. Then the question is, "Who's going to take over after I'm gone?" At least one of the kids tends to come back in their mid-thirties or so, and they realise that they quite like that lifestyle (beekeeping). I see this in operation after operation. Have you spoken to John Miller?

Steve: No

Randy: He's a very large beekeeper just down the hill. He's a well-known beekeeper in California and North Dakota, he's well-spoken and has been on TED talks, but he is a 5th generation beekeeper. None of his kids were interested, and it bugged him that it was going to die without passing the baton, then his son Jason came back and got involved. I told my sons here that we can go to the almonds and after that make it all up into nucs and sell the lot, and I'd have enough money to retire on, no problem. Not a lot of 30 year olds have a business offered to them, but I said that it was up to them, and they said they wanted to take it over. That's what they're doing.

Steve: They must have had something to do with the bees at some point?

Randy: Oh, they grew up riding on the seat of the truck, going to almonds, and for 25 years I ran the bees up to Nevada for the summer. They did that since they were infants so they were there but they never paid that much attention; it was more of a job that Dad did. Last spring we got the cheques from the almonds, over $200,000 from pollination, and I said, "Here you go," and I passed them the cheque book. I told them not to confuse cash flow with income.

Those cheques had to last them until this time next year, and expenses really eat them up. They got on it seriously. They worked extra hard because they didn't hire enough labour to replace me. They didn't realise until a few months in how much work I did! Later on they said, "Dad, we had no idea how much stuff you were doing behind the scenes, at night and in the mornings before we got there...my God!" After 50 years of experience I'm able to work very quickly and efficiently, so it would take more than one person to replace me.

So, my sons worked really hard. They didn't want to fail. This spring they went to the almonds with the most colonies we've ever taken there, and the strongest colonies, and after that they sold the most nucs that we've ever sold. They not only made the money stretch but they had a surplus at the end of it which they used to re-invest in the business, so I was very proud of them. They really did a good job.

Now what I'm enjoying is being a father with two sons that show up every day, and we all get along. I told them that I'm not interested in micro managing, running the business is up to them, but I will answer any questions that they have. There have only been two occasions, due to weather events, where I got in there and told them what I thought without being asked. I said, "You don't know this but around 25 years ago we had a weather sequence like this and it was a disaster for me. Here's how you can prevent disaster: stop what you are doing right now and shift to this

Learning from mistakes and high levels of rain

Steve: Wow, so what was that weather, drought or something?

Randy: No. I had a brilliant idea a long time ago, when I had 250 hives, to go to the almonds and stick a queen excluder in every hive. Later I could go back and find which box did not have the queen, put that on the top over a horizontal divider board, give them a queen cell and then immediately sell the boxes with the old queens as singles. I'd be left with 250 hives with newly-mated queens in the almonds.. I found a buyer and everything was great, I sold the 250 singles, but immediately afterward it rained for a month straight! None of my new queens could mate, and I was left with 250 mostly queenless hives that went laying worker. It was just a huge monster setback. It was one of those learning experiences that you never forget.

Steve: It probably doesn't rain for a month very often in these parts, does it?

Randy: Well, it's been like 25 years since that occurred, and the boys didn't have that experience. I said, "Stop splitting your hives right now, and hang on to enough for recovery in case the rain doesn't stop." The weather did change and we mated out 1,500 queens in one afternoon, right here. It was just incredible. I was watching the thermometer. It has to get to 70°F for a queen to fly out to mate, and it was 65, 66, 68, then at 1:30pm I heard the drones come pouring out of the hives. It was 68 degrees and they weren't going to wait anymore. I opened up a nuc and saw the queen with the mating sign – you know what that is?

Steve: Yes

Randy: You don't see that very often. I opened up the second nuc; there's a queen with the mating sign. I opened up the third nuc and the queen was just flying away to get mated, so I just said, "OK, that's enough." In that half hour the queens from 1,500 nucs were mated out. That was last year.

Steve: Do you know where the drone congregation areas are? It doesn't really matter I guess.

Randy: It doesn't really matter. I haven't found them here. At Ray's you can go out at 1:30pm and you can watch the drone comets right in front of your face. They are even at head level. I have photographs that I took where you can see the queen getting mated in mid-air and drones right there coming after her, all right in front of your face.

Steve: Fantastic

Randy: They don't even have to get far off the ground before that happens.

Steve: I must admit, in my country it's quite common to have rain and badly mated queens.

Randy: Oh yeah. I was over in Scotland, or was it Ireland, last year, and one of the beekeepers showed me a weather graph and it had never gotten up to mating temperature in the entire summer. Not one single time. It's very rough; you would have to have a locally adapted bee to mate at lower temperatures. Italians would not have mated.

Steve: No. The big beekeeper I met in Scotland finds that it's mainly the Carniolan type of bees that do best for him.

Randy: Yeah. A lot of them are going back to Apis mellifera mellifera there (Amm). It's even a question whether that was the native race. From what I'm hearing that may be a total misnomer, and that there was actually an English race of bees that was replaced by Amm. What I see from most of them who are very adamant about their Amm breeding programs, is that they may all be misinformed as to the true native race!

Steve: I haven't seen anything about that before

Randy: Not in print – there was a discussion on the Bee-L forum that the original English bee was brown, not black, and that an introduced remnant population exists in a country in northern South America.

What happens is that somebody with authority says something, and it gets repeated and repeated, and after it gets repeated enough times it gets in all the books, and everybody keeps on repeating something that was never supported by evidence in the first place. It's not FACT. I'm one of the myth busters out there. I get some strong push back but so far I haven't called things wrongly.

On being a scientist

Steve: Your articles in the American Bee Journal have a big following don't they?

Randy: If I wanted to be published in a scientific journal I would write for that, but prefer writing for a trade publication for beekeepers, who can really use practical information to apply to their management. The feedback I get from beekeepers across the country and the world is that they love the way I write. I intentionally write with fewer four syllable words, intentionally use beekeeper slang, and I mess my English up on purpose, so it sounds like a beekeeper talking naturally.

Steve: That's what they want isn't it? Cut to the chase, what do we need to know?

Randy: Exactly. What I hear is that I have become the de-facto extension agent for the country. An extension agent is hired by the government, the USDA, to translate science to the farmers. They read the science and then they get it out there. What's happened over the last few years is that most of the extension agents have retired and not been replaced. California's Eric Mussen, is a friend,

Checking magnetic hive entrance for discs

Bear damage

and one of the major ones, but he recently retired, and he was pretty well the last one of the old school. By default, I'm just kind of filling in until others step up. What gives me credibility with beekeepers is that I've got propolis on my hands – I actually make my living from running bees commercially. To other beekeepers this gives what I say great weight, since I'm not "just an academic" giving advice.

We have an informed discussion group for bee biologists, called Bee-L, and more than any other forum out there for beekeeping, in this one, if you put bullshit up there you will be corrected! Not destroyed, since there is no animosity, it's just that if you are going to say something, that you need to be able to back it up with evidence.

Anyway, around 2004/2005 when CCD gained traction there was a shortage of bees for pollination. I started pollinating in the 1980s at about $12 a colony and we had worked up to $45 a colony by the 2003/2004 pollination, but the next year when there was a shortage the growers bid the price up to $155. In one year it jumped from $45 to $155! That set a new baseline for almond rental rates, and has since profoundly changed our industry. I wrote to the editors of our major beekeeping magazines and told them that we just had a watershed event: the U.S. beekeeping industry now made more income from pollinating a single crop, than from all the honey produced the rest of the year. We'd now shifted from being mainly a honey-producing industry to a pollination services industry.

Bear damage

Dealing with bears and growing cannabis

Hey, good morning! It's my son Eric. This is Steve, from England. He's an author. He's going to be interviewing me today and tagging along.

[Conversation with Eric about what they are doing today]

Randy: How bad was the bear damage?

Eric: It's ongoing – I've been up there every day. There's two hives gone and a bunch of them were knocked over but I've fixed that. It's a really crafty bear. You should see the fence I've put up there; it's above and beyond anything I've ever done before.

Randy: Ah! Brain to brain against a bear, huh?

Eric: It's been digging underneath the neighbour's fence, then excavating under our hives – he's a digger, he's got pits everywhere. I put the wire right down into the hole and he just came back and dug deeper, he went underneath again. There's been no more damage this morning, but no signs that he's been around. All the bait is still there.

Steve: What's that about bait?

Randy: We hang strips of bacon on the top hot wire, so that the bear will get a shock directly to the nose or mouth. One good shock to that area will make a bear avoid the fence perhaps for the rest of its life.

I had three hives there with magnetic traps on the entrances – are they still there?

Eric: No, all gone.

Steve: What's all that about?

Randy: I'm running a large experiment to get hard data on the drifting of bees and varroa mites from hive to hive. We've glued steel tags to the backs of 6,000 bees, and we're recovering them with magnetic entrance traps when the bees drift to other hives – this yard is a mile away, yet we have recovered tags there.

We have bear issues here!

Steve: Do you poison them, or trap them or shoot them?

Randy: No, we put the electric fences up. You can shoot them if you get a permit but that's not too popular with the neighbours and we want to get along with our local wildlife, not hurt them. So, what we do is to train the resident bear to avoid our electric fence around the apiary, and the resident bear then keeps other bears away.

Steve: OK

Randy: We just make it painful for them. They learn that those white wires will give them pain. But this guy has learned how to get underneath them. We've got 55 bee yards around here, so that's a lot of maintenance. Similarly, we don't want to harm our local skunks (which pillage hives at night) – we also train them rather than kill them.

Steve: We don't have bears or skunks

Randy: Oh that's Ian over there – he's in our cannabis garden. They've grown it for years. That's our number one Ag. industry here, now it's legal.

Steve: OK. I wasn't expecting that!

Randy: Our county is known for producing high-quality cannabis – for both medical and recreational use. We've had a thriving quasi-legal industry for years.

My sons argued that they'd make far more money in the cannabis business than with beekeeping, and that the work was easier. I explained to them that cannabis wasn't much more difficult to grow than were tomatoes, and that the price would soon drop. Over the next five years the wholesale price went from $6,000/lb to around $600. So they are glad that they stuck with the bee business as their main income.

Steve: They did something with that in Canada recently didn't they?

Randy: Yeah. The price has plummeted over the last few years so it's not as lucrative as it once was. Anyway, I'll tell you what; we'll wander down to look at this trial I'm conducting, which is why I'm interested in that yard a mile away, which we will go and take a look at afterwards.

Steve: How long have you got before you boot me out?

Randy: You can hang around all day

Steve: OK

Randy: I'm going to multi-task. The only critical jobs I have today are to check on that bear yard, and also check on another yard where I have a patty trial going on. I'll explain all about it when we get there. My assistant Brooke will be here anytime.

Steve: So, you don't do one trial at a time?

Randy: No, we've done 5 trials this season.

Running trials to solve the three biggest challenges facing bees and their keepers

Randy: I have a list of questions that are important to beekeepers; they need the answers to scientific questions and a practical application. I try to find other researchers who might do it, so I write the proposal for them and have them submit it to an organisation to get funding, and I push for it to be funded. Then if they get to do it they get to publish a paper. My name is never mentioned but I'm the one that puts the whole thing together. If we have really important questions and I can't find anybody to do the research then I just step in and do it myself.

There are 3 major things that we needed in beekeeping. We needed a good model, to model varroa in the hive – the build-up and effects of various treatments. Nobody was doing that so I spent the entire year last year developing an accurate, user-friendly computer model for both beekeepers and researchers to use, and now it's online and free to use.

Steve: I know, I love it.

Randy: It's being used all over the world, so I checked that one off the list. The next one was that we needed a late summer treatment that was legal for varroa for when the honey supers are on. That's the oxalic acid in glycerol application method that I'm currently working with USDA to get approved by the EPA.

Steve: What does that entail?

Randy: EPA requires that we present a data package showing that the application method is safe for the applicator, doesn't cause adverse effects to the bees, and doesn't contaminate the honey. EPA won't require additional data to demonstrate efficacy at killing mites, but we are also running experiments to demonstrate that. So far, things are looking very good, and this new mite-control measure could be a game changer for our industry, especially since beekeepers are reporting that amitraz appears to be failing (as did the two previous synthetic miticides).

I personally stepped off the synthetic miticide treadmill in 2002, and have run a successful commercial operation since then using only the organically-approved biopesticides thymol, and formic and oxalic acids. I've also stepped up my selective breeding program for varroa-resistant bee stock – we now requeen our entire operation each season only with daughters whose mother colonies have managed to control varroa on their own for a full year. I'm "walking the walk" for the benefit of the bee industry – to keep track of the amount of labour involved, and my degree of success. If I'm successful, then it's likely that other large-scale queen producers will copy my suggested methods.

Also, we need to improve our pollen subs, so I have a big trial going on right now.

Steve: Is that just comparing different types?

Randy: Not comparing; I'm improving. I found out that there are two deficiencies in the best current ones, so I'm trying to correct that.

Now there's also a big question about late season immigration of mites into hives which is poorly understood. What I'm seeing is that there is an evolutionary process going on that people are just not aware of. We are creating a monster, this combination of varroa and Deformed Wing Virus (DWV), which is acting very differently to how varroa alone works on its natural host, Apis cerana. We need to understand this, and we need to quantify what's happening. I've been talking to others who are looking at this and nobody's moving fast enough, so that's what this trial is about.

The yard just down there is involved in that trial. Let's walk down and I'll show you.

Research and understanding varroa

Question: when colonies collapse from varroa, how many mites move from those colonies to other colonies? How far do those mites move? Is it mainly from certain colonies to certain other ones, or is it just widespread? Is it due just to simple drift, with the bees getting confused, or is it intentional drift with them going further afield, or is it to do with robbing from other colonies? Those are a bunch of questions that nobody can answer right now. This experiment is designed to try to answer all of those questions!

These hives with a "D" on them are donor hives, and the paint colours are to show what colour the bees in these hives are marked. These ones up here are receiver hives, completely free of mites, with magnetic entrance traps and stickyboards to see if they pick up drifted bees and mites. The donor hives have extra mites put in so that they will collapse.

Steve: Oh, gosh.

Randy: So, we have a dozen hives intended to collapse. The receiver hives all received two different types of synthetic miticides which are very effective (we haven't used synthetic miticides in our yards since 2001), so we cleared all of the mites out of them. Once they are cleared of mites you can put a sticky board underneath, and any mites found on that board must be from outside. We also have these hives on electronic scales so I am monitoring the hive weights. I can tell whether the hive is collapsing or being robbed out or whether it's robbing by the weight change.

Steve: What's the weight difference between collapsing and being robbed out?

Randy: If it's collapsing you get a minor weight loss due to the bees, if it's being robbed out you get a major weight loss due to the honey being taken. I'm doing this now when there is no honey flow so that any weight gained would have to be from robbing.

If one of these donor hives has a major weight loss at the same time as one of these receiver hives has a major weight gain, and I can tell that the bees from one hive ended up in the other, because they are marked, so I can see the mite drop would be from drifting. If on the other hand there is not an increase in the number of marked bees from donor hives, then the mite increase is due to robbing. OK? I can differentiate between drifting and robbing.

Steve: This is incredible stuff...

Randy: Each of these receiver hives has a sticky board which we check twice per week and count how many mites dropped. We monitor the donor hives with alcohol wash and as they are about to collapse we mark 500 bees. We use the magnetic traps to recover any tags that drift to the receiver hives. We know which marked bees came from which hives. There are also two control hives that have no mites, which are not collapsing, to compare the drift from them to the donor hives, which are collapsing.

To mark them we use this chair with a little table, and on the table we have some glue and some carbon dioxide canisters to anaesthetise the bees. We pick the bees out of the collapsing colony, drop them into a container, knock them out with a shot of CO2, then they are groggy for a while so that we can pick them up one at a time. We then put glue onto their back and attach a tiny coloured steel disk, and we do 500 of them. They go back into their hive. We have done 6,000 so far. We have to hand punch and hand paint every disc, and then manually attach them to the bees.

Then I also have, for hives within a mile radius of here, entrance traps to catch any drifting bees with discs on them. These are powerful magnetic strips which go along the hive entrance; if any bee goes into the hive it gets the disc ripped off its back. These are on all of my receiver colonies in this yard here, and also in hives 500 yards away, and half a mile away and a mile away. We are getting discs picked up at all of them!

Steve: Wow. These are workers as well, not drones. [stupid question]

Randy: Yes, there's no drones here at this time of the year

Steve: Of course :(

[Introduces me to Brooke & we move to a nearby shed]

Randy: Here's all the tags which we have to glue on the bees. Here are the sticky boards; Brooke counts the mites on here. We are going to be able to correlate hive weight, number of mites falling, the strength of the hive – we grade them for strength as they collapse, so I know how many bees each tagged bee represents proportionally – so this is going to be a monster data set.

Steve: You haven't got any hunches yet, from what you have seen so far?

Randy: Why would you say that? [laughs] Of course we have.

Steve: Are you able to share?

Randy: Urm..OK, we've got receivers right in the middle of the donor hives, we've got some up by the house here, we've got some higher up at the top of the property and some far away. Which ones would you expect the most drift to go to?

Steve: The nearer ones.

Randy: Yes, the obvious assumption would be the nearer ones…but it appears to be just the opposite! [laughs] We both find that very interesting. The further they get from the hive proportionately, the greater the amount of drift up to a point. Most of the drifting is apparently going to hives from 500 ft to a half mile away – compared to hives at 15 ft, 60 ft, or a full mile distant.

Steve: They are going from collapsing colonies?

Randy: Yep, and surprisingly, perhaps just as much from the healthy control hives – one control hive exhibited the least amount of drift, the other, more drift than any of the collapsing hives – go figure!

Steve: Maybe they are just trying to get the hell out of there, get as far away as possible.

Randy: Why? They are not trying to save their own lives, they are doing altruistic self-removal. A bee doesn't care about saving its own life because it can't reproduce, so there is no genetic or evolutionary reason for a worker to save its own life, that's why they are willing to sacrifice themselves when they sting. So, the big question is whether there is an innate behaviour for bees to drift to distant hives, or perhaps the DWV in the bees' brains affects their behaviour. We still have much to learn!

You've got a fan here [to Brooke – pointing out animal poop on the steps to the trailer in which Brooke performs her lab work]. That means you've got to pee on it to talk back to the animal. We'll turn our backs for a moment…

Brooke: Right [laughs]

Randy: I don't know if you realise it, but I intentionally pee all over the property on the trees and posts and stuff, so I'm always talking to the animals.

Brooke: I probably won't do that

Randy: I've got an advantage, because I can pee higher than you can. That's what they do; they want to see who pees the highest.

Brooke: Oh really, they can tell?

Randy: Yes. If you can pee higher you must be a bigger animal. It means, 'Wow, there's somebody bigger than me peeing out here, staking a claim."

Steve: The things that you learn…it's amazing

Brooke: Yeah, exactly; you have no idea! [laughs]

Randy: One major thing I've been trying to do is to breed for varroa resistance, and now I'm trying to show a way to do that for commercial queen breeders. I'm a big preacher about it. The article I just sent off was, "The future of varroa" and it's about the evolutionary direction varroa, deformed wing virus, and beekeeping are most likely to take over the next several years. Right now it's going in the wrong direction and I'm trying to help change that direction.

Steve: The direction is negative why? Because people are just treating all of their colonies, so you don't get the pressure to become resistant?

Randy: That's true but that's not why the direction is negative. The reason is that we take away the fitness cost to the parasites from killing their host, by rewarding the benefit of dispersal before the host colony dies – we eliminate the cost by restocking the deadouts with fresh colonies each spring.

But this dispersal, to be of benefit, must occur when it's warm enough for bees to be flying. I'm currently running an experiment with tagged bees in collapsing colonies to determine the amount of bee and mite drift that takes place late in the season.

Here look at this; this is one thing that's going to revolutionise bee breeding. [shows me a device with a wire loop for holding a cup of bees].

Steve: What is it?

Randy: It's for doing an alcohol wash – in 60 seconds you get a mite count. This little portable device plugs into your cigarette lighter in your truck. It's how we do all of our mite washing now. This invention has value because it makes monitoring varroa levels so much quicker and easier than any other method out there.

Steve: That was one of my questions as well, about mite washing. A lot of people either don't do it or they do a sample where they test say three colonies out of a yard, then they'll assume that it represents all of the others.

Randy: Last year we mite washed a thousand hives to start with, pretty much the whole operation. If somebody would have asked me two years ago, "What proportion of hives do you wash – do you wash all of your hives?" I would have said "No, that's crazy, nobody in the world is going to wash all of their hives!" This year my sons said, "Let's wash all of the hives."

We find it cost effective; it saves us money. We make more money, and it's more profitable, for us to wash all of the hives at one point in the year. It pays for itself just in terms of the amount of mite treatments that we can skip, and there are huge benefits – you remove all of your colonies that are going to crash, you can do the selective breeding program and you have a very good assessment of where your whole operation stands on mites at that time. What was it you were saying about Ray?

Steve: He was saying that he breeds from Saskatraz queens brought in from Canada.

Randy: Yes, they have been breeding for varroa resistance for a number of years, I have one of their queens over there.

Here's the deal: you can't slow me down. [Randy marches off with me struggling to keep up]

Steve: That's fine [I lied]

A bit about Citrus Greening Disease

Randy: If you want to talk to me, you've got to run.

I worked with the Israeli firm Beeologics, a cutting-edge gene tech company working on RNA and RNAi and so forth, looking for a cure for Israeli Acute Paralysis Virus. This was during the CCD (Colony Collapse Disorder) epidemic.

Then I got a call saying, "Will you work with me Randy on another RNAi project? We need you to do it in California and we need you to build a lab that will meet USDA insect containment standards – a quarantine lab." The idea was to be able to raise 5,000 of these little insects a week, and to rapidly propagate potatoes and tomatoes in a laboratory. I said, "What's the budget?" and was told it was very low, so I bought this junk trailer for $350 and parked it here on my land.

Steve: Wow

[Randy showed me around the trailer/lab, which had shut down and was now just a storage area]

Randy: I'm the guy who builds whatever he needs to have built, from scratch.

Steve: What were you trying to do with all these insects?

Randy: We were looking at citrus greening disease which is hitting all of the orange trees worldwide. Also, there is a similar insect and similar bacteria which is hitting tomatoes and potatoes – it's a major problem, so we used that one as a model for citrus greening.

We also did a project on mosquitoes down in Brazil, so I do research things independent of bee stuff. The idea was to create sterile male mosquitoes, which you can do by the way you feed them as larvae, then do a massive release of these sterile males to help to eliminate the mosquito population. They just had a press release a few days ago down there, so he's still going with that.

Future proofing – building a varroa resistant strain

I'm in this unique position. I don't write for grants or apply for funding; beekeepers just send me money and say, "Keep doing what you're doing." As far as I know I'm the only independently funded researcher where the industry I do my research for just sends me money to spend how I best see fit. I have a working budget, I can hire Brooke, I can pay for having all of those bees marked, and all they want me to do is just publish it so that they can use the findings. It's a pretty cool position to be in.

Steve: Yes, it is. But that wouldn't happen if you didn't deliver, would it?

Randy: You got it!

Steve: What about the fact that you've got the commercial beekeeping operation as well as research?

[In the truck, which was very noisy so sound quality of recording is terrible]

Steve: Is it difficult because the needs of your research are different? You have this yard and want to do some crazy research on it, but that might mess up your honey crop or something.

Randy: OK, so here's the deal: when I handed the boys all those cheques last year I told them that I'm handing it over on the condition that I always have 1,000 hives at my disposal for research, and that I will pay them if they lose money on it. I said I can make it worth their while, I have a research budget, so if we run a control group that doesn't get treated and they crash I'll pay the difference.

Steve: So, is that for all of your hives?

Randy: Well, we start off with about 2,000 in the spring, we take about 1,000 to the almonds the next winter and sell about 1,000 nucs after almonds. This year I used around 400 hives for research. For this feeding trial that we're going to this afternoon, Eric asked me how many of the 76 in that apiary are going to make it to the almonds. I said that of the four treatment groups three of them should do well but the control group probably won't make it to the almonds but I will make up the difference in lost income.

I also have manufacturers who have products that they want to sell to beekeepers. If you were a manufacturer, say you came up with this brilliant new "save the world medicine" for bees, and you go to beekeeping suppliers to let them know about it, pretty soon what you are going to hear is, "What does Randy Oliver think of it?" Anybody who's got something to sell wants to come and talk to me. I just had a group here from Israel and another bunch of guys from New Zealand.

What I need to know is what claims they are making about their product. I'm independent, I'm not a salesperson, and all I do is collect hard data to support or not support their claims. This year we tested a product which the manufacturer claimed would take care of varroa and nosema and all this, but the data did not support any of that, which (unfortunately) is often what happens.

Steve: It makes you wonder what kind of a person would go to all that effort if they didn't believe in it themselves?

Randy: They do. Not only that, but they put hundreds of thousands of dollars into these things, but they don't know how to run an actual trial to get good data. They give it out to beekeepers and ask them, "How does it work?" Beekeepers always say that they like it but that's not hard data. I get hard data and run controlled trials here and send it back. I have a spreadsheet for them to fill in where they say how many times they want me collect the data and it automatically calculates the cost for them, so I don't have to waste my time on giving them quotes.

I only do contract research if they agree to full transparency, so that I'm never put into the position of having to withhold information from beekeepers. All the donations from beekeepers don't quite cover the costs, labour, and my time for my research and writing. I'm incredibly grateful for, and appreciative for those donations, since they allow me to do what I love, but the main thing to me is that those donations tell me that what I'm doing is of benefit to the beekeeping community. That's what drives me – that I'm lucky enough to have found a niche in which I can benefit the beekeepers of the world.

Here's some of the receiver hives for that trial I'm running with the magnetically tagged bees in the varroa infested colonies. So far these pick up a higher proportion of drifting bees than the ones much closer to the collapsing colonies. We're probably about half of the way through the trial so we can't make any conclusions yet.

Steve: What about the robbing situation, is that happening? Can you tell?

Randy: So far, not strong. The data is coming in faster than I can keep up with at the moment. I've got way more ideas for research than I can possibly do. This trial started off with me using yellow bees and black bees, but it became apparent that the queens had been open mated and the workers from yellow queens were not always yellow, and the workers from black queens were not always black. That's when I decided to use the magnetic disk and retrieval method which had been used before, back in the 1970s on some research tracking bees in the almonds.

[Magnetic Retrieval of Ferrous Labels in a Capture-Recapture System for Honey Bees and Other Insects. Norman E. Gary

Journal of Economic Entomology, Volume 64, Issue 4, 1 August 1971, Pages 961–965, https://doi.org/10.1093/jee/64.4.961]

Randy: The main time that mites are transmitted from one colony to another is late in the season, when some are collapsing. Our beekeeping practices are creating an evolutionary pressure that changes the varroa mite from being a parasite into a parasitoid. That's what we need to be aware of and what we need to change.

Steve: How do you change that then?

Randy: Firstly, we need to prevent varroa from building up in our colonies.

Then we need to shift the genetics of the American honey bee population to naturally mite-resistant strains. Once we all run mite-resistant colonies, DWV will no longer be a problem.

Steve: So the main commercial queen breeders need a financial incentive to go in that direction?

Randy: Yeah. As soon as the first one actually has resistant queens on the market the buyer demand is going to shift.

[phones Brooke to ask for code to enter gated area]

This is it – this is the farm, right here. Eric, my son, has been in the battle of the brains with a bear here. Oh yeah, all new fence…OK, so the charge is good…look, you can see the damage. The bear dug out there, oh my God! Come here [laughs] – that's all from the bear, all that digging, all that dirt. A couple of days ago there was none of this here. Wow. I think Eric has met his match here with this bear!

Steve: Yeah.

[Randy looks at magnetic strip from a hive entrance]

Randy: Look, there's one red tag on here. We normally don't have dirt on it, but the bear did that. This is the second tag that we've recovered from a mile away.

Steve: Are bears nocturnal then?

Randy: Yes.

Steve: That's crazy. I would have thought that it would have knocked all of the hives over but it's just a couple.

Randy: No, they usually just start with a couple. I think this tag is red, but red and pink always confuse me, so I'll bring it back.

Steve: I'm hanging back because I look like a bear.

Randy: Yeah, these bees will be a little touchy because they've just been attacked.

[Bees start to get in my hair so I rapidly move back towards the truck. Somehow, I didn't get stung]

Steve: How long has Brooke been with you?

Randy: She did a little bit of volunteer work...actually she worked on that tomato project, then I started training her this spring. As you can imagine, working for me is not the easiest thing in the world!

[I laugh in agreement]

Randy: I have a zero-tolerance policy for any excuses. If something goes wrong, you don't even think about saying why; I don't give a shit.

Steve: Yep, just fix it.

Randy: And don't do it again. Luckily, she talked to my previous assistant, who told her, "Yeah, Randy's easy to work with if you don't make any mistakes at all." [laughs]

I challenge my people all the time to really understand and think about what they are doing and why. Anyway, I'm taking this tag back to confirm the colour.

Steve: That's quite a distance that it's come from, isn't it?

Randy: This is over a mile. We've had two tags here now. Each tag represents roughly 40 bees. So far we've recovered the equivalent of 80 bees drifting to those three hives with entrance magnets. Now, there's 15 hives there, which is 5x the number of receiver hives, so 80 x 5 = 400 bees. We have documented that 400 bees have drifted here. If every fourth bee is carrying a mite, that's 100 mites that moved from a mile away to here. That's pretty sobering, and it's what I'm seeing.

Steve: So much of your work has been on the varroa mites. Is that because it's "personal" because I think I read that you got wiped out in the past by mites?

Randy: Yes, it's personal, because I'm a beekeeper, but it's also practical. I've got some really strong emails going back and forth with other bee researchers. I tell them that what they are interested in doing may be of academic interest, but isn't of practical application to beekeepers. I try to steer researchers towards projects whose results may be directly of benefit to beekeepers.

Steve: I recently read about Samuel Ramsey finding that mites attack the fat bodies rather than the haemolymph.

Randy: Yeah, that's something that a number of us have suspected for many years.

Steve: I guess that explains the susceptibility to viruses, because it's part of the bees' defence isn't it?

Randy: Yeah, that's part of it. The thing that Sammy didn't talk about, which I pointed out to him (he was here – I get lots of visitors, plus I also helped get him funding for his next step) was that if there's a wound the fat body cells migrate towards that wound. That totally supports his whole thing, which is that all that the varroa mites have to do is make one wound and the fat bodies keep moving towards it.

Steve: Right

Randy: What we need to do is help other researchers so that we are all moving forward. You see, I don't have a career, I don't have to watch what I say, nor apply for grants, nor get papers published for career advancement; I can work strictly in the interests of beekeepers, which is pretty cool.

Steve: Those bees at the last place were definitely after me.

Randy: Well yes, when a bear comes in, you can imagine…

Steve: Yes [laughs]. I got away with it though.

Randy: Let's look at this tag properly in the light. It's red, definitely red.

Personal story - how it all began

Steve: Were you a researcher first then you came to bees or…

Randy: I grew up being a biologist. The biological interest was mine, but encouraged by my parents. My favourite book as a child was my fathers college biology text. And as a teen, "Culture Methods for Invertebrate Animals." My father was a well-loved teacher, and I grew up knowing the joy of sharing learned information with others. I was able to visit two biology teachers and talk with them.

My third-grade teacher came up to me and said, "What are you doing?" I explained that I had magnetised a needle and taken a thread out of my sock,

Field trial of supplementary bee feed

and bent it into an arm and made a compass. They said, "Do you want to show that to the rest of the class?" So, by third grade I was demonstrating science experiments to my class and in junior high school I was working on quite a few experiments, primarily on research of flatworms. You can train a flatworm, then if you cut it in half and let both halves grow back to full worms, the memory is retained in both. In high school my teacher hooked me up with the university, so I helped out with some research looking at immune responses in cockroaches. Unbeknown to me, I had been put into accelerated classes in high school.

Steve: This was aged 14 or 15 or something?!

Randy: Yeah. It was around then that I got my first honey bees. I learned about beekeeping in the traditional way from a side-line beekeeper.

Steve: Was that around here?

Randy: No, that was in Southern California – Newport Beach. My first honey bee experience was in hiving a swarm in a neighbour's yard into a cardboard apple box. My father had a student who'd mentioned that her father was a beekeeper. We got introduced, and I apprenticed to him. I had a strong back, and he had knowledge. He ran about 200 hives, and I learned the craft from him. I kept a couple of hives for the next years as a hobby.

I worked full-time as a short-order cook to earn a living, and to put myself through college. I got accepted to the brand-new University of California at

Irvine, where I soon gravitated to entomology, and wound up being asked by the Dean to establish a world-class insectary, mass-rearing insects for research. What an incredible experience!

By this time, I was working summers with my brother in construction – there was a housing boom going on, and I learned several trades.

After graduation from University, I got a chance to move to a remote cabin deep in the woods of the Sierra Nevada, in the old-time Gold Country (I of course took my two bee hives). I lived there with my young wife for two years – no electricity or running water, and a wood stove. Another wonderful learning experience. I quickly got the reputation of being The Answer Guy, since I not only had book learning, but could also build things, weld, do mechanics, etc.

Steve: Amazing

Randy: After a couple of years in the woods, I got the chance to move to a small town and run a farm store, where I studied up on animal husbandry, farming, and gardening. After a couple of years, I got the hunger to go back to school, and applied to Humboldt State to become a wildlife biologist. Instead, I got accepted into fisheries, since a professor needed someone who could study the crustaceans feeding the salmon fingerlings in an aquaculture system.

Although my partners in the farm store offered me a very attractive incentive to stay, I again loaded up the old pickup truck, and with only a few dollars in my pocket, headed off with my wife to look for a place to rent. I struck a deal with a homeowner to hand dig a septic system on the property, and again worked my way through graduate school.

By this time I was serious about education, since I'd left a good job to go back to school, and got straight A's. I was popular with the grad students, since I had construction skills to help them with their projects. I loved getting back into research – when one professor said that our grade would be based upon a research paper that couldn't be over ten pages long, I dug into the subject and wrote a 100-page paper on the history of every managed fish species in the State. The professor scolded me and said "I told you not more than 10 pages." I replied, "pick any ten." By this time I was loving teaching, and headed up study sessions for the other students to prepare for finals.

When I visited the University a couple of years after I graduated, I found that everyone knew my name, since the thesis writing class (which I never took) had been using my thesis as their example.

Steve: That must be really gratifying.

Randy: By the time I graduated, my brother said that things were booming in construction in SoCal, so I went back to make my fortune, with the thought that I'd move back up north and start an aquaculture business. We were better builders than businessmen, and didn't make much profit. So I moved back to the Gold Country, and found work in construction.

I took a university class on beekeeping from Dr Norm Gary at UC Davis, and asked about getting into a doctoral program in apiculture. He told me that at the time, the job prospects for someone with such a degree were slim. By then I had a different wife and a kid on the way, so I became a contractor, and built houses, learning all the trades. On the side, I started building up a small beekeeping business, reinvesting any profits back into the business (this was early 1980s).

When my sons were in grade school, I asked them what they did in class for science. They said that they read the science book. So I started volunteering to do hands-on science with the kids. Soon I was in great demand by other teachers. All this time I was active in environmental issues, helped our local land trust to get going, and kept self educating.

Steve: You just couldn't help yourself could you?

Randy: At this time the charter schools were taking off, and I got a job teaching 6th graders. It was wonderful to be able to enthuse young people with my style of real-life education. For example, as a Christmas fundraiser, I had my class create a beeswax candle making business, learning lessons about bookkeeping, jobs, packaging, and salesmanship – the kids loved it!

The next year I rented a room in a building and helped to set up an interactive science museum, with me starting the Nevada County Science Center. I specialized in hands-on science for charter school students, and quickly had a waiting list. I learned how to be an effective and beloved instructor by the school of hard knocks, and had parent after parent tell me that I was the most important teacher their children had ever had. This gave me a great sense of purpose. It also prepared me for something completely unexpected.

Steve: What was that?

Randy: I'd built up to around 250 hives and was taking them to almond pollination. And then came CCD, and the shortage of bees in the winter of 2004–2005. The new gold rush started when the almond growers bid the price for colony rent up from $45 to $155 in one year! Starry-eyed beekeepers loaded up semis with half-dead hives and hauled them to California without contracts, expecting to get rich. Some complained loudly when the brokers told them that their hives were worthless.

A beekeeping buddy called me and said, Randy, you need to write an article for ABJ (American Bee Journal) about the truth of the matter. It had never occurred to me to write for a magazine, but I researched and pounded out a 10,000-word article and sold it to ABJ in November of 2006. Joe Graham, the editor, got good feedback and asked me if I could write some more. As it happened, I was seriously starting to self-educate again at the time.

Varroa was kicking my butt, and it occurred to me that I, with a degree in entomology, was allowing a mite to get the best of me. By this time, we had the Internet, and I could gain access to research publications. So, I "went back to school."

After publishing a few articles, I was invited to speak at the Oregon State Beekeepers convention. The only convention that I'd attended up to that time was the California one, where I had been an unknown (although by that time I'd been president of my local club for some years). I took my current wife, Stephanie, along. When we got there we were shocked, as I was treated like a rock star. We couldn't believe it, and still laugh about it to this day.

Steve: Randy the rock star!

Randy: My life hasn't been the same since. I took a sabbatical from teaching science, and after a year of paying rent on the room to hold it, realized that I'd moved from teaching schoolkids to teaching beekeepers. I also started growing my operation, until I was running 400 hives by myself. I of course made every mistake imaginable, which is really important if you want to understand all the possible pitfalls in beekeeping.

Beekeeping, bee research, writing, and speaking engagements now consumed me. Twelve years later I look back and realize that the remodelling project on my house had come to a screeching halt – we still have bare insulation and electrical wires sticking out in the rooms that I started working on back then.

I had unwittingly stepped into a niche that was crying for someone with my life history – making a living at beekeeping, handy with my hands, with a scientific background, and the ability to write, with teaching skills learned in the classroom. I learned quickly that what beekeepers appreciated was straightforward, practical information, written in the language that they speak, not dumbed down nor pretentiously puffed up. But most of all, accurate and evidence based.

Steve: Yes, we do appreciate that

ScientificBeekeeping.com

Randy: I started the website ScientificBeekeeping, and before long was getting letters and invitations from all over the world. I was in the best of all worlds for me – doing what I loved and found exciting, not being restricted by any institution, so that I could say whatever I wanted, and do whatever I felt was right. I cannot thank the beekeeping community worldwide enough for their appreciation of my efforts to serve them – it's been my highest calling and honour to do so.

During this time, my sons Eric and Ian came on board, and we grew the operation to over 1,500 hives, without the use of any synthetic miticides. I felt that it was best to walk the walk and set an example of how to run a profitable and sustainable beekeeping operation in the real world.

Steve: Then you handed the business over to your sons

Randy: Yes. They grew up as infants riding in car seats in the bee truck as I hauled loads of hives to Nevada, and learned the ropes through experience. I'm proud to say that their first years at the helm have been quite successful. This now frees me up for full-time research and writing, and, Stephanie hopes, finishing up the house remodelling.

There was no plan. My wife and I keep shaking our heads thinking, "What the hell happened?!" It's just so bizarre. So that's where we are. I'm trying to claim my home life out of this. I would be speaking every single week of the year if I didn't turn down invitations.

[A bee stings me on my belly through my tee shirt]

Randy: These bees just don't like you!

Steve: Ow. Yeah, that one didn't like me. It's probably the wrong colour tee-shirt [dark brown/grey]

[I moved away, deciding to get my veil from the truck. With hindsight, it's possible that the bees from the "bear yard" had left some venom or alarm pheromone somewhere on me, or it could be hotel shower gel; who knows?]

Randy: They are a little touchy out there. They were just bumping off me too.

Steve: It's alright.

Randy: So, this has been a huge surprise to me

Steve: Your rock star status?

Randy: Yeah, I don't really like it. People want to take selfies with me which I'm not really comfortable with. It also means that I have to be really careful about what I do and say.

Steve: Because you get misquoted?

Randy: Right. I don't need to make money. I've already set up enough for retirement and I handed the business over to my sons. I've always been an educator and I'm somebody that wants to give back to the beekeeping community. That's what I do. I know how to be a commercial beekeeper and I can translate the science. I can talk with any researcher at any level and then write to the beekeepers in a way that they like. I enjoy doing research and enjoy helping out, doing something meaningful.

Steve: Would you write a book?

Randy: That's the plan, one day. It would be something like, "The Why of Beekeeping," – why you do what you do. My website is called scientificbeekeeping.com. Science is not about test tubes, it's about the logical steps you take, it's a way of thinking. Everything you do in beekeeping should line up with the biology of the honeybee. We need a good understanding of the biology of the honeybee and the varroa mite and so forth; to me that is what scientific beekeeping is.

Steve: Sounds good

Randy: It's about the science and the practical application of the science, step by step.

Persuading beekeepers to breed varroa resistant bees

Steve: So, the biggest threat is caused by the varroa mite, and the way we do beekeeping makes things worse, and the only way to reverse this is through the big queen breeders, right?

Randy: Yes, initially. They are big businesses and they won't make a change without a good reason, and the reason would be if consumers start to demand it. I'm trying to help with that. As soon as a queen producer can say, "I've got bees that don't need any mite treatments," everyone will be scrambling to catch up. I have set out a way for commercial queen breeders to do that. I'm a big picture guy. I'm looking at the whole thing and thinking, "What needs to be done? Not little band aids – how do we change the big picture for beekeepers worldwide?" I'm trying to fill in the holes to get that done.

[Get out of truck back at Randy's place]

Steve: It's like the Mediterranean, this climate.

Randy: Yeah, it is. We are on a banana belt right here. In the winter it will often be frost free here. We have fig trees. Oh look, my first ripe persimmon. I've got to take this to my wife. C'mon, we've got to split it three ways.

Steve: They are very sweet, aren't they?

Randy: Oh yes.

[Go into house and eat persimmon]

Randy: Hey Steph? Stephanie? Oh, she's not here. Have you had persimmon before?

Steve: I've had it before but not straight off the tree.

Randy: I'll give you a slice first, because some people don't like the texture.

[It was delicious]

Randy: I tell those guys, the big queen producers, I tell them that they will be out of work if they don't change. Do you like carrots?

Steve: Yes

[Randy starts scraping the skin off a carrot and hands it to me, then proceeds to make both of us a giant sandwich each]

Steve: I made my own queens by grafting larvae for the first time last season. It was only about 10 queens, but I was really pleased.

Randy: Good!

Steve: I hadn't realised just how small the larvae were that you are meant to graft from.

Randy: This spring I did most of the grafting. I graft 300 in an afternoon.

Steve: Do you use a Chinese grafting tool?

Randy: Once you get used to the Chinese grafting tool it's great. OK, let's go. [Charges out of house munching sandwich and heads to truck)

Colony Collapse Disorder

Randy: So the CCD epidemic started around 2003. Dave Hackenberg brought it to everyone's attention in 2006 or 2007, but it had been responsible for the shortage of bees for almonds in the winter of 2004–5. At this time fluvalinate and coumaphos were failing as miticides, DWV was rapidly evolving, a new form of EFB was running rampant, and the invasive wave of Nosema ceranae was spreading across the country. There were also some virulent forms of IAPV hitting some operations hard. It was a "perfect storm" of a number of things all hitting the bees at the same time, plus most everyone was running bee stock of very limited genetic diversity. CCD blew through by around 2009, and most losses since then appear to be mostly due to inadequate nutritional and varroa management.

I ran a field trial with Beeologics, in which we inoculated colonies with purified IAPV that I had cultured from a friend's operation that was collapsing from CCD. My friend Dr Eric Mussen was the Monitor. At the end of the trial, Eric and I sat on a couple of hives and marvelled at how in that yard on that day we had inspected hives at every stage of CCD collapse. We wondered about the steps leading to the dramatic and sudden collapse. At that moment, I started to formulate in my mind what had happened, and over the next several days realized that several positive feedback loops worked together to cause the collapse.

I drew up a model of the sequence, and published it [see http://scientificbeekeeping.com/sick-bees-part-2-a-model-of-colony-collapse/]. A couple of years later the leaders of the USDA lab studying CCD told me that they had immediately adopted my model as their working model to explain CCD – that made my day!

Steve: I remember you saying something about disease and pathogens?

Randy: You cannot transmit a disease; you can only transmit a pathogen. If the host does not resist the pathogen, then you've got a disease! People need to get their terminology right. I don't mind it when audiences get this wrong but scientists should know better. My previous assistant told Brooke, "With Randy you've got to learn how to speak differently, how to use your senses differently – but it's all going to be to your benefit." If you were to ask Brooke she'd say that she is learning to talk differently and to do her whole thought processes differently. It's because I have no time for all the bullshit and games that most people do.

Steve: Where's this then?

Randy: This is our nearest town, Grass Valley.

Steve: Oh, I'm staying here.

Randy: If you want to see something interesting go right here. This is the old gold mining place, the Empire State Gold Mining Museum, and they have some cool stuff – all the old machines that they used back in the day.

Steve: Yeah, I might have a look at that.

Thoughts on Manuka honey

Steve: What about Manuka honey? A load of bullshit or…?

Randy: OK, I was speaking about this the other day. I was showing somebody a chart of an economic bubble and how Manuka is right at the top of one such bubble. It's supported by loads of Chinese consumers who, right now, think that Manuka is the best thing ever. In truth, there is good evidence that Manuka honey is of benefit as a wound dressing, but that's it. There is not a shred of scientific evidence to support the belief that it's of special benefit when eaten, or in hair shampoo, and on a whim the whole industry could crash. They told me about what the New Zealand government was doing to support the Manuka honey industry. I said, "Meanwhile, over the water in Australia they are out-planting you with other kinds of leptospermum plants that you don't have." I would not be investing in Manuka in New Zealand right now! About a tenth of a percent is for medical use and the rest is just bullshit.

Steve: Yeah, that's pretty much the view of Peter Bray, who owns Airborne Honey. He was saying the same thing.

Randy: I will say that some of the health benefits of honey are incredible, and it does have antibacterial properties. It's just a marketing thing with Manuka though – they were the first to really push it.

Revenue streams

Steve: In your bee business, or maybe I should say now your sons' business, is it all about the almonds? Is that where the money is?

Randy: Half of it is from the almonds and nearly half is from the sale of nucs. There's a little bit from the sale of honey. We are still in the red after almonds and we're in the black after nuc sales, and it's just profit after that.

Steve: How many nucs do you sell?

Randy: We sold 1,000 last spring.

Steve: Who buys them?

Randy: It's split mostly between side-line beekeepers and hobbyists. We've never advertised but even last spring there was a demand for twice as many nucs as we could produce. It's not a bad business to be in.

Steve: How much would a nuc cost?

Randy: We sold them for $160. We are right in the middle price range; not the highest and not the lowest. We sell nothing but top quality nucs and we get really good feedback. Our business model is this: we sell 1,000 nucs which is 5,000 frames, and every spring we buy 5,000 frames, already assembled, ready to drop into a hive for $2 a piece. The next spring we sell five now perfectly-drawn combs covered with healthy bees and a new queen for $160 – that's a $28 margin per frame times 5,000. We have a free exchange policy – no questions asked. We are not claiming varroa resistance but we are getting a lot of positive feedback about that.

Dealing with swarming

Steve: When they come back from the almonds they are about to swarm, right?

Randy: Yes, so we split them all; we split every colony 4 ways.

Steve: So, you need to have made the queens by then?

Randy: Yes. We sell two of them and keep two. We start out with twice the number of hives that we need to take to the almonds each spring and we select the best, cull the worst and combine them, so what goes to the almonds is really good.

Steve: When is the first grafting for your queen rearing? Presumably just before they come back from the almonds?

Randy: Yes, that's right. We can make up 150 nucs in an afternoon.

Steve: Based on my experience of being in a car with commercial beekeepers [Mike Palmer & Richard Noel take note] they mostly terrify me with their driving, but so far I'm OK with you.

[We talked a bit about how Randy does his queen rearing but sound was bad. It's all here anyway: http://scientificbeekeeping.com/small-scale-queenrearing/]

Advice on going commerical

Steve: What about advice for somebody who wants to be a commercial beekeeper?

Randy: The worst thing you can do is start with 300 hives. If you think, "I've got some money, I'll buy 300 hives," then you are likely to lose everything. You need to start off much smaller and make your mistakes at a small scale first. A successful beekeeper can triple his number of hives each season by splitting them. If you start off with two hives and try to grow each year, and make the bees pay their own way, then you can see if it's for you. If you are not making a profit there is no sense in doing it.

To be successful you need strong healthy colonies, so anything weak has to go. That needs to be your number one focus, and the second thing is timing; you've got to be in the right place at the right time. You also need to do some marketing – selling your products (honey, beeswax, nucs, etc) and services (pollination, bee removals, etc). All those things like trucks and equipment are expenses that don't directly make you income, so I would focus on what makes you money and as that grows you can invest in more stuff. Healthy strong colonies create wealth out of thin air. You can start out with the cheapest equipment you can find and if the bees are making you money then you can build from there.

The beekeeping year

Steve: In your operation, after the almonds hopefully you get some honey later on, but then you get a dearth don't you, in July/August?

Randy: Yes.

Steve: What do you do then, feed the hell out of them?

Randy: No, you just don't take the honey off.

Steve: Right.

Randy: We do some feeding some years but we don't focus on the honey crop so they should be able to use their own stores.

[Crossed a cattle grid]

Steve: Is that for horses?

Randy: It's for cattle; it's called a cattle guard – they won't cross it.

Steve: We have them for horses where I grew up in the New Forest.

The media only seem to want to report about bee losses don't they? They only want to tell the story that honey bees are all dying. It's strange isn't it?

Randy: Absolutely. I see that when people are fundraising they just propagate these untruths. It's what you'd expect from the corporations. I find it disgusting that an environmental group that I support would engage in this kind of fear mongering, but they do.

Right, so we are arriving at a yard with some weak hives that need feeding to get them up to strength for the almonds.

Steve: Is this part of your trial, or are these just normal patties?

Randy: No. This was a late yard of nucs. These are now all singles but most of our hives at this time of year are on doubles. Even if they stay as singles over the winter this will be a reservoir of young queens. Queen age is nothing to do with chronology; it's how many eggs they have laid, so with these colonies being small the queens are still young. It can help with recovering any winter losses. All you have to do is put a box on top of these in the spring and those queens will take off like gangbusters. I'm going to ask Eric what his plan is. I know he's got more colonies now than he needs to clear the almond contracts so these would be reserves.

We have three different tiers of contracts. One tier is for premium colonies (12 frame or better) which you get top dollar for, around $215.

Steve: Is that 12 frames of bees?

Randy: Covered with bees, yes. Then we have contracts for 8 frame colonies and contracts for 6 frame colonies. These here could actually be used to fill the 6 frame colony contracts; you don't make the top dollar but they could still go to the almonds if Eric wants that.

Steve: The almond growers are desperate for bees aren't they. You are perfectly positioned here…

Randy: Our trucking expenses are much less than many people further away. We can winter our colonies outdoors too, there's no need for indoor wintering, and we can spread them out so they don't have too much competition.

Steve: How cold would it get in winter – say the worst? Down to zero or something? [I meant Celsius!]

Randy: No, no, no. Down here you don't hit 20°F. Up at my place you might just hit 20°F [-7°C], I think the record low is something like 15°F. Normally it's above freezing. We do get some snow but then the temperature comes up. It's rare for them to go more than 10 days without having at least one flight.

What we've found is that feeding the pollen sub here, at this time of the year, makes all the difference in the world. After we've got varroa way down and we've got Deformed Wing Virus cleared out we feed them pollen subs and syrup, and we get these rounds of brood before winter. Those emerging bees are going to be your winter bees so you go into winter with a strong mite- and virus-free cluster. These are tiny colonies right now.

Due to our nectar and pollen dearth here in late summer and fall, for 25 years I hauled all my hives over the summit of the Sierras to irrigated alfalfa (lucerne) in the state of Nevada. The honey flow and fall forage was much better there. But once the boys got families of their own, it became a lifestyle issue, and we learned to make up for the lack of natural forage by feeding high-quality pollen sub. We get our bees into optimal health by November, and they're ready to roll come February.

In the trial that we are going to see in a minute, I did the same thing; I started with small colonies with the intention of building them up big. Some are doing it and some are not. Last week when I checked there was some natural pollen coming in from somewhere, which screws up the trial. I'm hoping I don't see any fresh pollen today when we get out there.

Increased CO_2 and the impact on the nutritional needs of bees

Steve: Ray was showing me that he uses pollen sub with some natural pollen mixed in.

Randy: We are using a very similar formula from the same manufacturer, who we are working with. I had theirs analysed and I told him, "Hey, you are deficient in two nutrients!"

Steve: Is that amino acids?

Randy: No, I'm way beyond that. One is a sterol and the other is a trace mineral, zinc. I told them they had zero of this sterol and they only had 20ppm of zinc when you should have about 75ppm.

Steve: How do you know that you are right? Is it because you have looked at what's in real pollen?

Randy: Let's get going and we'll talk on the way. There was a researcher, Elton Herbert, Jr, at the USDA who was just top notch and he devoted his life to some fantastic research. He died unexpectedly aged 40. I'm picking up his research because nobody else did. He was working on sterols that bees need. In mammals we make our own cholesterol from scratch; insects don't, so they have to get it from their diet. Each different insect species has different critical sterols. The critical sterol for the honey bee is 24-Methylenecholesterol.

[see here: https://www.sciencedirect.com/science/article/pii/0022191080901353]

Some pollens have a lot of this and some don't.

There is research to show that the nurse bees concentrate the 24-methylene-cholesterol from the pollen brought in and then they put it into the jelly which they feed to the larvae. When there's no 24-methylenecholesterol in the pollen the nurse bees will steal it from their bodies and pass it on. I looked for it in the analysis of the pollen sub and there was absolutely zero.

There was an eye-opening paper that showed our elevation of CO2 in the atmosphere was causing a decrease in the amount of protein in goldenrod pollen – as had been shown in crop species.

[Ziska LH, Pettis JS, Edwards J, Hancock JE, Tomecek MB, Clark A, Dukes JS, Loladze I, Polley HW. 2016 Rising atmospheric CO2 is reducing the protein concentration of a floral pollen source essential for North American bees. Proc. R. Soc. B 283: 20160414. http://dx.doi.org/10.1098/rspb.2016.0414]

If protein is decreasing, so would zinc, iron, and other trace elements. In various pollens, zinc appears to be a limiting nutrient for honey bees – there's not enough to take full advantage of the full amount of protein in the pollen. I had the standard pollen sub analysed, and it's zinc content appeared to be less than a third of ideal.

The major expense in the pollen sub is protein, but if it is missing critical nutrients like 24-methylenecholesterol and zinc then perhaps half of it is being wasted. By including all critical nutrients the manufacturers could probably drop the pollen content and reduce overall costs, while giving the bees exactly what they need.

Steve: They could probably charge more too, because it would be the only one containing all of the essential nutrients.

Randy: Right. I got the manufacturer to make up a custom batch for my field trial. I have four groups of hives with different patties on them. One is the standard commercially available pollen sub, one is the custom batch including 24-methylenecholesterol and zinc, the other is just a sugar patty containing exactly the same amount of sugar that is in the pollen subs (negative control), and finally natural pollen mixed with sugar which has exactly the same protein content as the pollen subs (positive control). I expect that the natural pollen will be best, the sugar patty worst, and the pollen subs somewhere in between.

Steve: You have 55 yards nearby. Do you have any others?

Randy: No, that's it.

Randy: The major colony broker at the almonds told me that our colonies were the best he'd seen all year. I haven't used synthetic miticides since 2001. The other guys with crappy bees are using them, which means they risk contaminating honey, so get no sympathy from me.

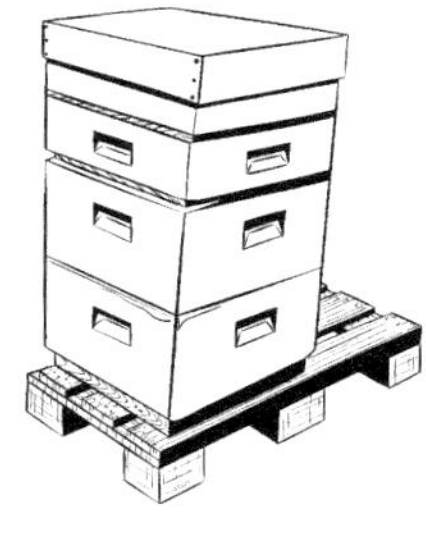

Steve: Is that amitraz or something?

Randy: Yes.

Steve: That's just started to be advertised in the

beekeeping magazines in the UK in the last year or so. Most people seem to use thymol and oxalic acid. What were we talking about? [Jet lag kicking in]

Randy: Erm...beekeepers not knowing how to keep bees well. Can I ask you to grab that plastic handle and move the wire so that I can drive through? If you grab the white wire it will be very painful!

Steve: Sure [I get out of truck & do the honours]

[Randy is inspecting hives, smoking bees and talking to me]

Randy: In the USA once you are a commercial beekeeper you are on a first name basis with the different manufacturers, OK? It's really nice. Mann Lake have their catalogue and their customer service and phone numbers and everything, then they have the line for commercial beekeepers, and they greet you by name; "Whatever you need." There's also a certain benefit to me being me...

Steve: Because you're "Randy Oliver"?

Randy: Yeah, but before anyone knew me they still treated me well.

Steve: Ah, you've got some wasps. I'm glad it's not just me.

Randy: What I'm curious about with these hives is whether or not they have natural pollen coming in. See the letters on the hives? They indicate the type of patty being used.

Steve: So, what's "T"?

Randy: That means "Test".

Steve: That's your new formula?

Randy: Yes. I had them add blue food colouring to it which has turned the patties green. This one has still got enough patty for a few more days, so we're not in a rush to add more. In this trial I am using an Amitraz strip to control it, because I don't want thymol or oxalic to have an effect on nutrition. Oh shit, look at that.

Steve: Yeah, pollen.

Randy: They're bringing in natural pollen. Fuck!

Steve: Does that ruin it?

Randy: It ruins it for this period of time, but that pollen will stop. For a short period of time we'll have pollen coming in. I wish that we didn't but I can't move them because that causes problems too. Look in this one – more natural pollen again. Fuck. Where is that coming from?

Steve: Looking around you can't see anything can you?

Randy: No, even if you walk out across these fields all you can see is green. There's a little tiny bit of clover but that's not what's coming in. I'm not going to spend time trying to figure out what it is. Here's the standard patty; see how the colour is different?

Steve: Yeah

Randy: It's an orange colour because they didn't put food colour in this one. They've got natural pollen too. So, if we have a warm winter with nothing blooming that will be good for this trial.

Steve: Will you just keep feeding them?

Randy: Oh, yeah. I'll show you the other patties.

Steve: Is it normal for you to keep feeding throughout the winter to keep them strong for the almonds?

Randy: No, only for a short time if they need it, but then they will go into their winter cluster.

Here's a sugar patty; they've usually gobbled that thing up in a couple of days. This colony is not looking big at all. That's ideally what I would want to see.

Steve: Yeah, because it proves that sugar alone isn't enough.

Randy: Well you don't prove anything in science like you can in Math. You can support a hypothesis but you don't ever prove anything. OK, so they are healthy but much smaller. If only this were consistent I'd be happy, but I'm not seeing that across all of these hives. The one that says "P" is for "Pollen".

[Back in the truck]

Randy: The thing that is very evident in beekeeping is that you don't know what you don't know. I mean, any idiot can blog about something. There are all of these inexperienced people telling other beekeepers what they should do and it really concerns me. It's unfortunate; there's a lot of shit out there.

Steve: I'm probably one of them! [Laughs] Although I tend to try to just pass on what I learn from meeting people like you. One thing I have noticed about all experienced beekeepers is that they always say, "In my area…" because they know that what works for them in their area may not be right for somewhere else.

Mistakes I have made

Steve: Have you got any examples of something that you've done that is really dumb?

Randy: There was that time I told you about when I failed to get all those queens mated

Steve: Oh yes.

Randy: Now I am more cautious, I don't go all-in on something. I tend to think things through, several steps ahead. Most people maybe think about one step ahead but I'm typically four steps ahead. That hopefully eliminates most of the dumb things that I could do.

Steve: Presumably that wasn't always the case though. Did it develop over time?

Randy: No, I think that's just the way your brain is wired.

Steve: Does that mean you are good at chess?!

Randy: I never really got into that so I don't know.

To be a successful beekeeper you need to be in a good environment and you also need to know what you want to do. If you are a queen producer or sell package bees that will require a different approach to something else. Maybe you want to produce specialist honey; everyone has to figure out what they want from their beekeeping so that they can get the most out of it.

The more you work with bees the more you can understand their cues, and when you understand them it can look like magic. What I've seen generally is that the two most important things for bees are varroa management and nutrition. If you focus on varroa management and nutrition then 90% of your problems disappear. Most beekeeper complaints are because they haven't paid attention to those two things.

Finding Apiaries and Global Warming

Steve: When getting apiaries do you just find a good spot and then approach the land owner?

Randy: We find land owners approach us. Bees are so popular now. We have way more offers to put bees on land than we can take up. With climate change we are seeing huge changes here in California, it's really noticeable.

Steve: Really?

Randy: Thirty years ago every colony I owned would go broodless around the first of November. Nowadays that just doesn't happen. The whole flora has changed; yards that used to be consistently on a certain flow at a certain time are not like that now. This right here – when these trees die they will not grow back. Because of climate change a new species of tree will take over. When the foresters replant after a fire they don't use seeds from these trees, they plant seeds from trees 500 feet in elevation down the hill.

Steve: So, it's happening fast?

Randy: It's happening very, very fast. In some areas you see it more than others and California is definitely one area where you see it. That's one reason why California is so anti-Trump.

Steve: I'm pretty sure it's not just California! [laughs]

Randy: Yeah, pretty much the rest of the world. It's unbelievable.

Steve: In my country "Liberal" means something different; we have Labour on the left, Conservatives on the right and Liberals are in the middle. That's what I am, bang in the middle; I like business but I think we need a conscience and a social responsibility. The thing is, Liberals just got destroyed at the last election, there's hardly any left. I'm in the same boat as you really; there's nobody in power who speaks for me.

Randy: If you focus always on being "anti" that gets you nowhere; you've got to be moving towards something positive. A lot of environmentalists are all "anti" things and I think they should be more "pro" something than always against.

Steve: Like Elon Musk with his electric cars?

Randy: Yeah, that would help, and a carbon tax which rewards people for their behaviour. It should go back to individuals rather than companies, like a cheque at the end of the year or a tax refund.

Find out what you like and what you're good at and focus your business on that. If you are going to be in business do something that you enjoy doing and that you are good at doing. There are a lot of different ways to get an income stream from beekeeping and you need to find out what your joy is and focus on that part.

Steve: Having seen how hard the work is I wouldn't try to be a commercial beekeeper but I like keeping bees, and I wouldn't mind having up to fifty odd colonies, but not much more than that.

Randy: I talk to a lot of beekeepers over in Europe; with a hundred colonies you can make a living in many countries. In France there's government support and in other countries the price of honey is so high that with a hundred colonies you can make a living. I used to run 400 colonies on the side, when I had a full-time job, you know? With one quarter of those you could make a living in many countries! That's a pretty easy living, only running a hundred hives.

Steve: Yes, when you put it that way…

Randy: Here that's not the case. You're talking five hundred to a thousand hives to make a living, depending on where in the country you are. It's easier for us to make a living here with the almond pollination and the earlier start to the season, but in other areas like the mid-West they get huge honey crops – they can take six months off each year. They can put the bees in cold storage over winter or sell them off in the fall then take their wife down to Florida, and they can make a living that way. They can start up the next season with new package bees if they sold them off. I know England is a bit tougher with your seasons – you are not in the best beekeeping area.

Steve: No, very wet. Nucs are very expensive. They are going for £200–£250, which is $300 or something.

Randy: Wow! So if you are going to focus on nuc production, and you dial that in scientifically to get the max out of those nucs, and you breed bees and manage the whole operation around that, you'd probably do well. We manage our whole operation around almonds; all year long we are thinking about what we need to do to get ready for almonds. It's not a "get rich" business but it's an honest life and we very much enjoy it.

Steve: The thing that I've most enjoyed so far, I think, is a little tiny bit of grafting and making my own queens.

Randy: We love that too.

Steve: It's magical. I just think, I've only got 16 colonies and even if I had 50, how do I know I've got a good queen because I'm competing with people who have thousands.

Randy: Not enough to compare to.

Steve: Mike Palmer says it doesn't matter, and that if you breed queens from your best stock then your whole stock will improve.

Randy: Yes, but you need enough of a gene pool that you are not going to get inbred. You are going to want to breed off 25–30 queens a season, so what's your selective pressure? We breed off of only about 25 queens out of 1,000, so we are putting pretty strong selective pressure on, but if you are only going from 50 down to 25 that's not very strong selective pressure.

Queen breeding

Steve: The biggest beekeeper in the UK is called Murray McGregor. He's based in Scotland and he has three and a half thousand and his season is all geared towards the heather. That grows on the hills and blooms in August so everything is geared towards moving them up onto the hills in time. It means that they come down quite late so he has to treat late and feed late to get them through winter. Another thing he does is he takes his own queens to Northern

Italy and they get bred from there, so the next season he collects packages with queens from them, but he gets them earlier in the season because it's warmer there.

Randy: I'm a big proponent of local regional queen production and queen breeding. It's like in Alberta and Manitoba in Canada. When Canada closed the border to package bees from the U.S., Alberta bought in packages from New Zealand, but Manitoba decided to be self-sufficient. When I talk to Manitoba beekeepers I ask them why they bother making honey; they could make more money making nucs and selling them to Alberta. The thing is, they shifted the paradigm. You don't do queen rearing in the spring; you do it in the summer then over-winter them in nucs to be sold next season, it's just a different paradigm. People are too limited in their thinking. It's the guys that think outside of the box that do well.

Steve: Do you have any of those funny little mating nucs?

Randy: Yeah, you'll see some on the left as you drive out. Eric bought a few hundred this year. I can't stand them but Eric wants to try them out. I used some Langstroth deeps split three ways last year and I said, "Look at these. We mated out the same number of queens, no odd equipment, no styrofoam, and it's faster to find the queens and pull them than it is in those tiny boxes." It takes forever to find the queen in those small boxes.

Steve: Probably less risk of them absconding...

Randy: Well, they starve in those small boxes. In a box with a full frame of honey you don't have that problem, but he's going to give it a go.

Steve: What are you doing tomorrow in Mississippi?

Randy: They've got a State convention and I'm speaking there a couple of times. They gave me a large donation previously for research, then asked me later if I could speak at their event, so I can't say no.

Steve: It's a long way away. I imagine it's a whole different country down there.

Randy: Oh, it certainly is. It's Trump country for a start.

Steve: But the plants and everything?

Randy: Oh yeah, and the humidity. It's hot and humid.

Steve: Does that mean small hive beetles?

Randy: Oh yes, lots of small hive beetles.

Steve: You don't get that up here?

Randy: We get a little tiny bit but it's so dry here that they can't reproduce well so they tend to disappear

Steve: Ray was saying it was bad in Hawaii

Randy: Oh yeah, it was pretty ugly when they first got there.

Steve: The main thing in the UK now is this paranoia about Asian Hornets

Randy: Yeah

Steve: It's obviously going to be a problem but I don't think it's going to be as bad as many think, based on what I saw in France. The further South you go the worse it gets, and we are cold and wet! Then again, as you say, climate change is happening.

Randy: Yep. You guys could get a really cold climate with the shutdown of the Gulf Stream.

Steve: Hope not.

Randy: You guys will have to build igloos

The fact that we have a President that is not addressing this huge problem of what humans are doing to the biosphere is just so discouraging.

Afterword

That was an interesting day and an interesting interview! Randy has a great life from what I saw; he loves where he lives and what he does, he is making a difference and having fun along the way. I suppose the only down side to being Randy Oliver is that there is an insatiable demand for his time, so he is always in a rush and always too busy. That being the case, I'm very grateful for the time he gave me for our interview. I did have to chase him about with my microphone arm outstretched, legs pumping to keep up (not unlike a Benny Hill comedy sketch) but I learned a lot. He is a clever guy who is able to combine science and beekeeping in order to educate beekeepers across the world, and it was a pleasure spending time with him.

David Kemp, Nottinghamshire, UK

Adam realised that in beekeeping; hives, extracting equipment, had all reached its peak, there was nowhere else to go apart from changing tin plate to stainless steel. So the future was breeding.

Introduction to David

I met David Kemp in the summer of 2017 shortly after returning from my trip to Vermont. At the time my vision was beginning to fail due to clouding of the capsules in my eyes in which the lenses sit. On a glorious August morning, I set off from my Manchester home in my old but very much-loved Porsche Cayman Sport for what I hoped would be an enjoyable blast over the Pennines. Unfortunately, poor eyesight and heavy traffic conspired to limit my fun, which was perhaps a good thing as I lived to tell the tale and avoided speeding fines. A year later I traded the car in as part exchange for a new Land Rover which is a much more appropriate vehicle for a beekeeper.

David arrived at the appointed time. As a disciple of Karl Kehrle, the famous Brother Adam of Buckfast Abbey, I would have expected no less. We went indoors to discuss his long life with bees. His advancing years had not diminished his enthusiasm or energy, so I was in for a treat. As you will see from the interview text, David did not always answer my questions directly, and often he would meander down various tangents along memory lane. That was fine with

Buckfast home apiary

me. Despite working for many more years as a bee inspector than as an assistant at Buckfast, it was apparent that his decade with Brother Adam had made a deep and lasting impression on Mr Kemp.

I was keen to get my hands on some of David's old photographs from his Buckfast days. A few weeks after the interview I was contacted by Andy Wattam, one-time National Bee inspector, who arranged to send me some images. We chatted briefly about David, who had been Andy's boss a long while ago. It became apparent that David was held in high esteem by many in the beekeeping and farming community. Andy explained how gentle David was when working bees, which is a trait I have found in all good beekeepers. By moving slowly and working calmly, without squashing bees, the task of hive inspections is generally a pleasant and stress-free experience.

Hive Inspection

Talking with David

How have things changed?

David: Things have changed. I did a programme called "The Bee Inspector" on Radio 4. Did you ever hear it?

Steve: I remembered that you mentioned it and I found a link to it on the internet

David: Yes, it's still about. I did a recording of it and it was sold all over the place. When I was working I had people contact me from New Zealand. Someone said that when they heard it it reminded him of his time in England with his father as a beekeeper – it's amazing.

David: There's one problem with the slides [photographs]. I had a word with a friend of mine, Andy Wattam, and he did them electronically. He said he can get them on a memory stick and send them to you. Are you OK with that?

Steve: Yes, that's fine.

Buckfast Abbey

David: They're now 52 years old and they're rare

Steve: They're delicate too

David: They're delicate. I started in '64 down there [Buckfast Abbey] and I had a Hillman camera, and one of the monks, Brother Ignatious, used to get me AGFA films. When they came out he would say, "What sort of camera have you?" [does German accent]. I showed it to him and he went, "Huh!" because of the Leica you see, the German Leica, and he couldn't understand how this cheap £10 camera could take such good images. But it was all about having a steady hand and taking them in the right light and that sort of thing.

I took them over '64 to '65, and it was a sequence of the work that we did

Steve: So a proper planned thing, for teaching or whatever?

David: Yes. I suppose at the time I didn't really think what the future held, but they were very good for presentations and stuff like that. Nobody had seen them before. Even local associations who could have gone to Buckfast to take photos – Brother Adam would have welcomed them – never did it. I suppose they thought it was always going to be there, but it wasn't, and now it's all gone.

Steve: Yeah. Apparently the mating apiary is just sat there rotting.

David: Yes, terrible.

Steve: I'm wondering if I can pop over there and grab a couple, I wonder if they would mind. Maybe I can buy a couple of mating nucs off them for posterity or something.

David: Have you ever heard of a man named Simon Croson?

Steve: I might have, not sure

David: He was a flight lieutenant in the Air Force, then he retired and took on beekeeping, and was a very good photographer. When I was doing a presentation he was taking shots of my slides! He produced one of their hives from that site. I was astonished to see it. There was a hut on the site, and he said it was stacked full of these hives. Vandalism had taken place, it was terrible.

I think it happened when there was a change in one of the Abbots at Buckfast, who wasn't bothered about beekeeping. They retired Brother Adam. I saw him in the August and he died in the September, he was a very frail old man. They had put him in a home.

Steve: What year was that?

David: Oh, goodness me, I don't know, maybe twenty years ago. [it was 1996 actually]

Steve: How old was he when you worked there?

David: Well, when he died I think he was 98, so he must have been close on retirement age, because he was quite frail then. [he was aged 66 to 76 when David: worked there]. But he was a tenacious German and he kept on and on and on. One of the reasons I left was that I thought, "this guy is going to die in his boots!" [laughs], which he did do virtually.

Things were changing so much. I was on agricultural wages, living as a paying guest with some grand people, and you couldn't see the future, you know? The apiaries: there were 320 honey producing colonies, the home apiary had 47 including the breeders for queen rearing...

Steve: Right

David: ...and the smallest apiary was about 28 hives I think, at Great Ambrook. The other apiaries were Dean, Dean Prior – they were about thirty something each, then there was Brent, Harberton, Ashburton that all had 40 hives on

Steve: OK

David: And then Gulliford, which had about 30 something in there. You see, agriculture had changed. In Devon in those days, there was no oilseed rape, and there was meadow after meadow, as Brother Adam told me as you passed the gates it was white...

Steve: What, clover?

David: Clover. Wild white clover. It was grazed by Devon ruby red cattle and longwool sheep, and as it's chewed off, the plants come back, and you could keep 40 hives in an apiary quite easily. But it did get to the stage that the amount of honey was spread over 40 hives when it should have been 20, and I said to Father Richard that the apiaries should be split up and cover a wider area. I knew Brother Adam wouldn't do that while he was alive.

Steve: It's continued hasn't it – that erosion of meadow?

David: Yes. When I came back in spring of '74 I got a job straight away as a bee inspector for the ministry, and oilseed rape came in. The local beekeepers were just inundated with honey, swarms of bees everywhere...

Steve: They didn't know what hit them

David: No! The forage was so immense then, you could have 60 hives or more in an apiary. Now, I don't really know.

But yeah, there was a system at Buckfast that Adam had worked over the years to suit the climate of Devon, which has a very high rainfall, about 365 inches a year [laughs]

Steve: Similar to Manchester I imagine

Brother Adam's queen rearing exploits

David: That's right. He had a system worked out over the years to his queen rearing, because that was what he was brilliant at. He understood the genetics of bees. Have you heard of Gregor Mendel, the Austrian monk, with his experiments on crossing sweet peas?

Steve: Like working out if you will get blue eyes or brown?

David: That's right, and Gregor Mendel kept bees. But you see, he couldn't control the matings.

Out apiary

David canning honey

Honey extraction

Steve: No

David: I think a professor in Germany told Adam to read Gregor Mendel so that he would see how the crosses of various things work. In the early days Adam went there [Buckfast] as a young boy, about 9 I think, and he was put on the building work. He told me his mother said, "Would you like to go to England to be educated in a Benedictine monastery?" At 9 years old he wouldn't know. There were quite a number of them came over at the same time, and he was put on this building work – he was quite a frail lad – and he became ill, with the weight of the work, cutting stone. So they put him in the kitchens. Brother Columban was the beekeeper there with a small number of hives, all different types, and to Adam's German mind, he thought this hotch potch mixture was no good at all, and he took an interest in the bees. Brother Columban saw that.

Steve: Yes

David: That was the way the monastery worked. If they saw somebody had a gift they could spot it at an early age and push them at it, which is not a bad thing

Steve: No, not at all

David: Also somebody gave him some rabbits. Rabbits breed like rabbits.

Steve: They do

David: I've seen a photo of him with all these different coloured rabbits around him, he's got one in his arms. He thought about how it all worked, these different colours being passed down.

It all began to form in his mind with the bees. They had at Buckfast then roughly "English bees" as you might say, black bees, and they also imported bees. They started importing bees long before he was there, from Italy. Italy at that time had some very good bee breeders, really excellent. Vast numbers of bees, mating stations and all that. They weren't bright yellow bees, they were a sort of tan colour.

A disease called acarine came along – you've probably heard of that, and the black bees went down like flies. The ones that survived were the natural crosses between the Italians and the black bees. Adam thought, "How's this?" It was the resistance of the Italian bee to the acarine disease. He started breeding, and he had several lines of the Buckfast bee, because if you just have one line it's no good; you've got to have several lines and keep crossing them.

You can get inbreeding, the Americans have had that. You went over to America and their breeding stock was very small, but once you'd crossed them – 4 way hybrids – they were tremendous colonies. There were two sorts;

Starline which were very light coloured, and Midnight which were very dark. But they were quite narrow in the breeding.

Steve: Right

David: Adam realised that in beekeeping; hives, extracting equipment, had all reached its peak, there was nowhere else to go apart from changing from tin plate to stainless steel. So the future was breeding.

I mean, they're still inventing beehives in England for goodness sake. The world standard beehive is the Langstroth and Adam thought at the time that he would go onto Langstroths, but then again, it was not a square hive, and as a German, you see, he would have had a square one. The frame depth was not deep enough for him to assess the brood pattern so he went onto the larger hive which is a Dadant, a modified Dadant.

Steve: Quite heavy I imagine?

David: Well, when you're young and fit, I mean I could lift them up – only as a brood chamber. To lift a brood box with a super or two supers you needed two people. In fact as Adam got older I used to nip across and take his supers off for him, and put them back on afterwards, because he was getting to be a frail old man.

He bought these hives over from the States, from Dadant…er, actually I think it was A.I Root

Steve: I've heard of them

David: Yes. They went through a machine and they used to stamp them. Inside you could see "A.I.Root"

Steve: I think they had something to do with starting package bees in America?

David: They could have done. So he ordered these hives from the States, because they were far cheaper than what was made in England.

Steve: That was quite radical wasn't it?

David: Oh yeah. I think he took over aged about 21 years, and he built it up to this excellent breeding station. We did do some artificial insemination there but when you start dabbling too close…you become like God. You see Nature – who knows how…well, the drones selected by nature have got to be absolutely fit powerful drones, and who knows what takes place? They say very often now that virgin queens mate with more than one drone, but we used to go to that mating station, and there were six drone colonies there, and you could see the queens coming back to their hive with a mating sign on the tip of their abdomen.

Buckfast Abbey 1960s

Steve: Yes

David: Now, until that was removed they couldn't mate again. My theory is that if they were successfully mated the first time they didn't go out again. Adam said in queen rearing, whatever you're doing in queen rearing, you must follow the bees.

[Note: research has now clearly shown that queens mate with multiple drones on the same mating flight]

He had a small queen rearing house with his breeders nearby. They could be worth, I suppose, looking at race horses, livestock, they could be worth several thousand pounds each to the man who knew how to reproduce them. Because every egg that the queen laid could be turned into another queen.

He walked Dartmoor and found this site that was ideal, and started stocking it with bees there. He used to take the virgins there. He'd have about six drone colonies. He'd start with twelve drone colonies, and selection, selection…I said "How do you finally select the ones?". It's intuition. When you work down from twelve to six, you start weeding out, until you get the best six. So he knew the drone line, which was the Buckfast bee, and then he crossed virgins from Greece, Anatolians, and so on, and he crossed them and saw the results. He was a very practical man.

Queen rearing

Steve: So he would take a Greek queen and get it mated by Buckfast drones?

David: Yes, and then he would select from them, because they don't all turn out the same, you see.

Steve: No

David: It's a mixture. He'd got that brilliant gift. He travelled abroad – he went for thousands of miles. He did go to the States whilst I was there, but he'd been to France, Austria, Germany, Turkey, Greece – he went to all those countries – Syria, you couldn't go there now. And he said to me towards the end that if anyone traced his steps they would not find the same bees, because governments were putting money into beekeeping, and people were buying in bees from outside.

Steve: Right

David: He went to very remote areas, to the ordinary beekeeper that maybe kept three or six hives, or twelve or whatever. He would select from them, where the stock had not been changed for years.

Keeping the queens alive and work with Dr Butler

Steve: So these queens being brought back from far off places, how on earth did he keep them alive?

David: He had special cages made out of balsa wood. Balsa wood for going abroad and ordinary pine for England. They had quite a large area for attendant worker bees, then a secondary chamber for candy. They have a similar principle now, but his cages were quite thin. He had an arrangement with the Ministry of Agriculture. He sent them back, the queens would be taken out and put in nucleus colonies on Dartmoor, and the worker bees were sent to Rothamsted to be checked for any form of disease, you see?

Steve: Right. So when you say "sent them back" they would go by post or something?

David: Yes, by air mail.

Steve: I don't suppose there was anyone else doing that at the time

David: No, I don't think so. When they imported them from Italy in the late 1800s I don't know how they travelled. There was a huge amount of trade, you know, it's always gone on with these countries in Europe. It may have been somebody coming back, and they would bring them in a cage, on his person or whatever. I don't know when postage started really, in its early days, but anyway they went to Rothamsted. Dr Butler [discovered 'queen substance'] , who died recently aged 102, was a great pal of Adams.

Artificial insemination and Brother Adam

When I went over to Rothamsted to learn how to collect semen from drones, Dr Butler was there in this little office with everything falling in on him from the shelves [laughs]. The man who actually taught me was called Ginger Walsh who'd been at Buckfast Abbey for a brief spell. Adam was going to do some artificial insemination. A virgin queen, when she's artificially inseminated, doesn't know she's mated, because she's inseminated under an anaesthetic.

Steve: Right

David: So you put the queen to sleep, you inject her with the semen from drones – it's like micro surgery really – I was taught by a German, who was sponsored by I think Bosch, and he could pop queens like lightning. He couldn't speak English and I had very little German, so he just showed me. I sat and watched him, then I had a go and he put his hands on mine and within a couple of hours I had it. It was very delicate work.

Steve: What about the queen not knowing she's mated?

David: Well, what you do is a day after artificially inseminating the queen you put her back to sleep again, and when she wakes up in a warm cabinet she knows she's been mated, for some reason. If you put her in the hive without that second anaesthesia she goes crazy. Whether they ever found out why the second anaesthesia works, I don't know. But after that, you put her in the hive and she starts laying.

You can speed up the generations by artificial insemination. But I thought, and I think Adam came to the same conclusion, that natural mating was the best. We put these very well-fed drones on the mating station on Dartmoor. They only fly at certain temperatures, but on a really nice day when the drones were flying it was fantastic, because you can hear the volume, and see the queens flying off.

There was an area called Foxtor Mires, I think it was mentioned in Sherlock Holmes, Hound of the Baskervilles, which was North of the apiary. Dr Butler did experiments with an artificial queen substance fastened to a twelve-foot fly rod to see how far he could draw drones away, and that's how far they went. He said there were loads of drones at this Foxtor Mires place. It was a natural mating area. He even had a dinghy and sailed out off the coast to see how far drones would follow.

Steve: Wow

David: Yeah. A lot of nature goes by scent, far keener than what we know. And that's how Adam worked with Dr Butler.

But crossing bees, it was amazing. I was reading last night about the Sahariensis – bees from the Sahara – they couldn't stand the English winter, but you could cross them to get tremendous colonies.

Steve: OK

David: They would go to the moors in July and would produce a huge amount of heather honey but they couldn't ripen it. So the cappings used to weep, and ferment. That was just a trait of this Sahariensis bee. Yet other strains were quite OK.

Developing disease resistant bees

Steve: So, the reason he became fascinated with the breeding side of things was disease resistance. When I was talking to Mike Palmer in Vermont it was obvious that he is obsessed with the quality of his queens, because it's a business and his livelihood. But with Brother Adam was it more just fascination with the science or was there a commercial side to it?

David: Well, he had to sort of make the department pay. Remember he was under the protection of a Benedictine monastery that financed all his trips abroad, and things like that. I think the worst honey season when I was there was 1.5 tons from 320 colonies, and the best season was 24 tons. As he said, in farming, with livestock, it's got to be about quality. You get quality bees, then you get the colonies to the right pitch at the right time for the honey flows, wherever you are, and then the rest is in the lap of the Gods. If the weather turns bad…

Steve: Yes. You do your bit and hopefully nature does its bit

David: Exactly, it will do as it wishes, and now we've got climate change. There is climate change going on. I used to compare notes with a beekeeper in Devon and the forage used to be a month behind, but now it's catching up.

Steve: I've just seen ivy today, not quite in flower but not far off. It seems quite early.

David: Yes it is this year. Some years you get a tremendous flow from the ivy. It's not the best for wintering on…

Steve: Because it goes hard

David: It goes hard.

Steve: But it's better than nothing

David: Yes, better than nothing. Adam used to take the hives up to the heather, but one year he didn't because the weather was bad and everything was against it, no rain in June and that sort of thing. They had to be supplemented. With the honey coming off in July they would weigh the hives and feed sugar syrup if they were too light. It was sugar syrup in those days, not the modern equivalents that you can buy ready mixed.

Steve: Oh, the inverted syrup

David: Yes, various things. But after coming back from the moors they would be picked up and weighed on scales, and if they needed it they would be fed using a large tray feeder. They lifted the hive up and because they knew the weight of the hive they could work out what stores were needed. The honey was taken off on the moors. We used to go up with a team of men on a lorry. The beekeepers would take the supers off the hives – they'd been left on bee escapes over the weekend – and we stacked the supers up on the lorry, and moved on to the next apiary.

He was brilliant at organisation, was Adam. It was spot on, typical German.

How I started at Buckfast Abbey

Steve: Were you one of many helping out or...

David: No. When I first went there the advert said "Beekeeping Assistant required for Buckfast Abbey" and I'd kept bees since I was 9 years old, and had this fascination about how bees worked. I had dabbled by buying bees from France and the Isle of Wight from Douglas Roberts, and could see the crosses. Douglas Roberts' bees were fantastic, not only were they quiet but they used to bring a lot of honey in. The French bees were vicious.

Steve: Were they?

David: Oh...they do well, but sting? When I was a gamekeeper I had some French bees, and my Labrador came along and got stung all round his lips and ears. He left me for the first time, he went back to the house

Steve: Can't blame him really

David: Whenever I went to the beehives after that he would stay back about 25 yards away. But the French bees I had were nasty. You could deal with them on a very good day but the slightest indication of rain or thunder or anything like that...and if they were confined for a long period they would just take it out on the beekeeper.

Whilst I was at Buckfast you never wore gloves. No suits like people wear now because they weren't about.

Steve: Just a veil?

David: I had an African Rifles hat from the second world war and a black net veil, and an apron. The tape of the apron held your veil down, and the apron protected you from getting messed up with propolis and honey.

But going back to the staff, when I arrived there and met Brother Adam for the first time, one Saturday morning, he came over with his hands drawn into his sleeves and his hood up...he looked like something out of MacBeth. He took me down to the bee department where Brother Pascal was working, who was also an excellent beekeeper – he'd been on the bees for 25 years – he was really good...

Steve: Yeah

David: So there was Brother Adam, Brother Pascal and myself who worked on the bees. Brother Bernard did the mail and stuff like that; posting honey off for Christmas – it used to go to Fortnum & Masons and a couple of stores in London, and a lot used to go privately in small boxes to various people. So we ran along like that for quite a number of years.

Steve: So you were in quite a privileged position

David: Yes, and looking back, how do these things happen? Why did I apply for a job at Buckfast Abbey? Although gamekeeping, which I was on for the previous six years, I could see that was all going to change. All the shooting was going to money. When I applied for the beekeeping job one of the old gamekeepers said it was the best thing I'd done and that all the shooting was going over to money. Because at Welbeck all the Duke's guests were Lords, Earls, Chief Constable, Prime Minister, you know, and now – everybody. If you've got the money you can shoot.

Steve: That's where you were before then, Welbeck? [Welbeck Estate, Sherwood Forest]

David: Yes.

Steve: Is that near here?

David: Yes, not far away. It's in the Dukeries. It's like Chatsworth, Thoresby – they say the Dukes had the land up here because it was sandy and it was good for game shooting.

But Brother Pascal became ill, he had stomach problems and lifting problems, so he came off the bees and we had various other people come, some of them just for brief spells, to learn. One chap came back, Peter Donovan, who'd been there as a refugee from London in the War. He then became old enough for National Service so he went off and never came back, until I think he lost his job. He worked for somebody who had tugs on the River Thames, and the chap sold the business or he died, and Peter was left in limbo, so he came back to work at Buckfast. He was still there when I left.

Steve: Right

David: He'd been back about two years before I left. He was there a long time, but he's dead now. Yes, a lot of changes…

Steve: What was the working environment like? Was it quite disciplined, friendly or…?

David: Oh yes, very friendly. Of course in those days if you had to be at work for 8am you were there for 8am, you know? Adam was a stickler for that.

Steve: So what if you were a bit late, would he tell you off?

David: Well, he wouldn't tell you off, but you weren't late, you weren't late. If you learnt beekeeping with Brother Adam you never forget, you just never forget, it becomes so ingrained and natural. We had one chap from a university ask Adam something, some question or other, and he asked it again in a fortnight's time, and Adam said, "I told you that a fortnight ago!"

A lot of days in the summer of course, we were out all day, and we had lunch out there, which was something, I can tell you. They had a fire box with a Primus stove in it. It was my job to see that the Primus stove was polished up to the nines and had paraffin in it. Then they had a food box, and in there was plates, cutlery and the food. In the back of the van there'd be a 28lbs tin with a bottle of the finest champagne honey mead in it. We found out that with three of us one bottle wasn't enough, so we had to have two bottles.

Adam used to make this honey mead. It was as good as a bottle of champagne, I can assure you; it was made under the same principles.

Steve: Yeah

David: All the best wines have monastic origins

Steve: They know their wine, don't they?

David: Yes. So I first met Brother Adam on the Saturday, and on the Monday, it was a warm April day, he said "I think we'll go out today." There were 2 smokers, 2 hive tools, and I had to smoke for Brother Adam. Smoking is an art – you don't choke them with loads of smoke – it's just a gentle swirl.

He taught me that when you've got an operator you just watch his hands and keep the bees calm while he's working. A week later there were 3 smokers and 3 hive tools in the van, and we'd go to an apiary. For instance, if there were 30 hives in an apiary Adam would do ten, Pascal would do ten and I would do ten. The following week we'd do a different ten, so every third week we'd seen all the bees, so you've got three pairs of eyes.

Steve: Right

David: But the thing with Adam was, when you got back to the van he'd say, "how was number 165?" and you'd have to remember [laughs]. When you work with bees, they're fascinating, you switch off and you look at three or four combs, and from that you should know what that colony is doing. If you take out a comb and the queen's there in full lay, quite happy, there's no point going through the whole lot, disturbing the bees, you close it up. You may come to a colony that wants a little more attention, so you've got that time that you stacked up [by not going right through the previous colonies].

Steve: I was going to ask about that, actually. I don't know if it's taught or it just takes time, but being able to just pop open a colony and know what's going on...how quickly do you know that there's something wrong?

David: Well, you get a feeling. One apiary can be doing differently from another one just up the road, because of slight local mini climates and things like that, and of course with Adam and his breeding, the queens in that apiary were not all from one type of bee. There could be a dozen Buckfast crossed with Greek, some Buckfast crossed with Anatolian, and so many others, plus one of the basic Buckfast main lines, so he could compare them. It was pure observation all the time.

Steve: So if you're looking at bees all the time and you are used to what a good colony looks like then it's easy to spot when it's not a good colony, so something must be wrong?

David: Yes. As Adam said, with 320 hives they are never all the same. If they were all from the same breeder there will always be some better than others.

The beekeeping year - winter

In the winter we worked indoors doing maintenance on the hives. He was absolutely keen on painting hives, keeping them up to snuff, and everything working right. The first outdoor work would be to change the floorboards on all 320 hives. You'd wait until the bees had a cleansing flight after the winter, then we go out in the van, 40 floorboards. We had a sequence; the new floorboard went to the side, then two people would lift the hive and place it on the new floor, so the old floor could be taken away. Then they would lift it up again, with the floor, and put it back in its place. You weren't dancing around to see which way it faced because they were kept two to a stand, four to a group, with entrances North, South, East, West, roughly, to prevent drifting.

Steve: Yep

David: And they were all numbered, they all had an enamel number plate on. If you lifted the hive up and the floorboard was clean...I used to get a pencil and mark down the number of the hive. When we got back to the van with all the old floors Adam would say, "Well?" and I would say "Hive number so and so Brother, clean." He'd get his apiary book out and record it. It was all relating to the queen and the crossing.

Or you'd get one with a lot of dead bees on it, you see. As you lifted them up Pascal or another experienced beekeeper could judge the weight, so if one came up light and there was another one very heavy, we would take a comb from the very heavy one and give it to the light one. When you're changing combs like that you have to be careful with diseases.

The second operation was taking out the old combs. He used to take three combs, sometimes four but mainly three, out of the brood box every spring. We used to place those combs in the colony the previous June. We'd go to the colony to select the old combs. He used to run 10 frames plus a division board in a twelve-frame box, and he'd put one old comb next to the division board and two on the far side.

Steve: OK, so the previous summer you moved the old combs to the edges.

David: He said that if there was going to be any mould in the hive over winter it would be on those outside combs. We would pull them out after winter, take them back and cut out the old wax – one of the filthiest jobs in beekeeping. Then we would have just over 1,000 combs to render down. It would take a day to render them down. I've seen Adam render 1,000 combs on his own, but he had the steam boiler, the steam pot. He'd put 36 in at a time and just bring it to the boil so it disintegrated the wax. The wire and debris would fall to the bottom and the wax would rise to the top. We used to pour the debris onto the floor and then put it through again, you would still get more wax, but it wasn't worth doing a third time. All the debris, the old larval skins and pollen, used to go down to the gardeners compost heap. You had to move it while it was hot and steamy because if it set is was difficult.

So your colony then went down from ten combs to seven. The brood frames were treated in caustic soda – you wouldn't be allowed to do that now – you can do it with washing soda now, but caustic...if it got on you it would burn

Steve: Yeah

David: Then they had to be rinsed, because the nails would rot with the caustic, then they were stacked and dried, the whole workshop was a drying room.

Steve: Then you would have to put foundation in

David: Yes. If the frames were needed in April, to bring the colonies back up to ten frames, we'd put foundation in around March time. Not too early, because the aroma goes off the wax, you know. And we used to re-fit them.

Steve: So did you have to sit there wiring frames?

David: No. The wax blocks, which used to weigh about 100–110lbs, were weighed, wrapped in polythene paper then put in sacks, with a couple of ears sown on so you could lift them. They were sent from Buckfast to the railway station, I think they used to go to Newton Abbot. Then they'd go by rail from

Newton Abbot to London, and they'd go onboard a ship to Chrysler Brothers in Ontario...

Steve: Wow. OK.

David: Now Chrysler, is it a German name? I'm not sure. But it was a contact that Adam knew. His wax that he sent would be milled into foundation and come back. It had vertical wires, crimped, and it was beautiful foundation. It came back in boxes with a layer of tissue paper, then a layer of foundation, then a layer of tissue paper and so on. Beautiful stuff. They could get it done in those days cheaper than it could be done in England.

Steve: And better

David: And better.

Steve: So that was already wired

David: Yes, and the wax was held in the frames, not by a wedge, it was a saw cut. The frames were very substantial to make them last longer, and they were spaced by two boot studs, so they only touched on that metal stud. There was nothing there for the bees to propolise, and he selected bees that didn't use too much propolis because it hindered you. You can select for it. The Buckfast strain use a very limited amount of propolis.

In the winter the queen excluders were cleaned with a warm paint brush and vaseline, on the woodwork, which prevents the bees sticking stuff to it.

It was all thought out and had taken years to get to that stage, but it worked, it ran smoothly.

Equalising colonies

What he'd do, he would equalise bees. This business about some forging ahead more and getting stronger. We used to equalise by taking combs of hatching young bees from the strongest colonies. They'd go in a brood box and be taken to the next apiary, and if anything needed a bit of help, you'd give it a comb of bees.

Steve: A comb with emerging brood on it?

David: Yeah, that's right. Adam would say that you can only equalise colonies for about three weeks, because the strongest would always forge ahead, but we would help the weak ones, you see. It was only to assist with management. If you go to an apiary and there's one hive very weak and one strong...it upsets the management, the operations that you are going to do.

When putting foundation in, with a good flow from the dandelions – and it was a good flow – he'd go three times in a week, he wouldn't put two combs in [of foundation]. He said that if the weather changes they won't draw the wax the same, so we'd put one in, then we'd go round again three days later and put in another one. It was a few minutes work on a hive, just open, take the feeder off and put in the comb. If the weather was bad he'd give them a slight feed, but only what they needed, dependent on the weather at the time and what was in flower.

Steve: Was that a top feeder?

David: Yep. The Buckfast feeder was the same dimensions as the hive, about two and a half inches deep, and they were waterproofed. First of all they were stained, this 'burnt sienna' was his favourite stain, and then they were waterproofed with paraffin wax. That dipping would preserve the wood and get them completely waterproof.

His beehives were bottom bee space. A normal Dadant is top bee space like a Langstroth

Steve: Oh?

David: They had bottom bee space because he said that when you put a screen on to move them it stopped the frames from rocking around.

Steve: They were doing a lot of that with going up to the heather

David: He moved them only once a year to the heather, you see, then they'd come back. But odd colonies like the selected drone colonies moved more. He would bring them in near the back of the greenhouses, then he'd put drone comb in there. He would look every morning to see when the queen had laid eggs, and then he knew that in 24 days they'd be hatching drones. You had to take them up to Sherberton before they started mingling with the home apiaries. At the right time these 6 colonies, selected from 12, would go up to the mating station. They would go up with 12 brood frames, plenty of food, a super, but no queen excluder. He would check them every time he was up there.

That mating station up on Dartmoor was an expensive station to run, because it was over 10 miles out on the moor.

The nuclei, on half Dadant frames, were four frames to a colony, and they would winter on either four half Dadants or eight (if you pull one of the division boards out). That's a good testing ground, you see, because Dartmoor was very severe in the winter, in those days. I don't know so much now, I don't know how much snow they get now. In bad winters the hives would be covered, there would be six foot of snow up there

Steve: Yeah

David: After it stopped snowing, after a few days, you would see a hole in the snow from the warmth of the hives. It was a testing ground for wintering qualities.

Steve: Would that be late August time, setting nucs up for winter?

Queen rearing – planning, preparation and crafty exporting

David: Well the queen rearing was in June, because he used to take the temperature at the home apiary every day, maximum and minimum, and every time he went to the mating station he did the same...I do the same now with a calendar – every day I write down the weather conditions and temperature. If you look back at the records it tells you, the fluctuations and what happens each month. He'd worked out that June was the ideal month for queen rearing.

It used to be: floorboards changed, old combs taken out, then as they built up new combs were added then supers were put on, then he went on to queen rearing. He used to raise about 600–700 queens a year with only 320 hives. There were 100 nuclei permanently at the mating station that he wintered there. They all had to be de-queened ready for the queen cells to go in, and they had to be equalised.

Steve: Were the queens taken out for putting into production colonies?

David: Yes, and he'd sell some. His turnover with queens was immense. If the queen was good we used to send it to Mr Harding for ten bob a piece, or if they were bad they had their necks wrung, or if they were mediocre they'd go to Rothamsted. Dr Bailey or Mr Harding, or the chop.

You have to be ruthless because if a queen is not performing right... sometimes they'd fight and would get a stiff leg so she could not perform, so she'd have to come out and another one put in. In normal re-queening, the young queen that had over-wintered went into the production colony and the old queen went back to the nuc on Dartmoor, to keep it ticking over until the next queen cells. It was changing all the time.

Steve: Yes

David: And he was sending queens to Charles Mraz of Vermont

Steve: That name rings a bell. He was really famous in Vermont wasn't he?

David: Yes, he's dead now.

Steve: I think he was one of Mike Palmers heroes.

David: Well Charles Mraz wrote back to Brother Adam and said, "send as many queens as you can!" They were a superior strain, you see. But it was illegal to send bees from England to the States, because of acarine disease. I don't think there was any acarine in the States. What happened was they cut the corners off these manilla envelopes and the queens were inside in their balsa wood cages, and when they arrived at the post office some bright girl would see the address and hear the buzzing inside, and they would be sent back to the Abbey.

Steve: Oh no. But normally you got away with it?

David: Yes. Also you could take drone pupae, lift them out, put them in a gelatine capsule, then into a money belt and go to London, express train from Devon to London, and you'd get somebody there from America to put the money belt on to keep it warm. Or you could send eggs; some comb with eggs in, because eggs are quite tough and it's 3 days before they hatch. They used to go express air, you see, then they could put them into a colony and then graft the larvae.

When we grafted a larva it was 12 hours old

Steve: Which is almost the same size as an egg, is it?

David: Very small, very small.

Steve: I can't do that because I can't see properly.

[brief pause to order sandwiches]

My childhood and how I became interested in bees

Steve: You know you said you started when you were nine years old? How did that come about?

David: In the village the school caretaker kept bees, and his house was at the bottom of the school gardens. Mr Chinrey his name was, and we used to see him with his brown smock coat on and his hat. When we should have been looking at the blackboard we were looking out the window at the beekeeping, because it was fascinating. He kept his bees in WBC hives.

Steve: OK

David: When he had his honey crop we used to buy odd pots off him – two and sixpence, half a crown a pound. Nowadays that would be eight pounds of honey for a pound (£). If he saw you were interested you could help him extract the honey, in a little 3 frame extractor. You'd go home, have your tea, then go and help him extract honey, and your reward was a pound of honey.

Steve: It would be worth it, wouldn't it?

David: Oh yeah, it was pure clover honey, almost pure, and I've never forgotten the flavour of it – fantastic.

Steve: I don't know if I've ever tasted it

David: It's a great honey. It was the premium one in the British Isles. We used to have gardening as part of the lessons, and one of the teachers thought it would be a good idea to have a hive in the school garden. They scraped up money to buy timber, because timber was short after the war, and we built a WBC hive from scratch, from drawings. We built it in the woodworking department, painted it and set it up in the school garden.

Steve: What school was that?

David: That was what we called Python Hill, a primary school, and I used to go there five and a half days a week. We used to go back on a Wednesday night for woodwork. The woodwork teacher Mr Trope was brilliant, an ex military man who took up teaching after leaving the forces. In the village, there were about seven beekeepers in the village of Rainworth, and the best one, who died before I met him was Mr Jack Radford, who was an engineman at Rufford Colliery. He used to operate the winding gear taking men up and down the pit shaft…

Steve: OK

David: …which was quite an exacting job. He had an apiary of about 36 hives and he paid his rates and paid for his car off the proceeds of that apiary. They were all in WBC hives.

I lived on North Avenue, we were part of a square: Kirklington Road, Python Hill Road, North Avenue and Cross Drive. Every house had a garden, except the ones at the corners, which had a space. There were air raid shelters there in the war. One of the gardens had a hive of bees in it and I used to sneak through the hedge and watch these bees. It was almost diagonally across from us on Kirklington Road, and they belonged to a man called Mr Lewin.

I couldn't afford a WBC hive at £5, that's what they were in those days, so I bought the floorboard and three lifts, and the frames to go in that. I made the roof myself and made the innards myself. When I got it set up my father painted it brilliant white. I used to look out my bedroom window and see this hive at the bottom of the garden, and I was itching to get some bees in it.

Mr Lewin worked at the colliery. I came home one day and my mum said, "Mr Lewin's brought you some bees."

Steve: Wow. Had he already put them in?

David: He had. The frames were there with foundation and he had put them in. He had also put a jar on it of thick syrup, with muslin over the top and that fed them. That's how people worked in those days; somebody's interested, some boy, they help them.

Steve: Yeah, that's nice.

David: Everyone said about National hives, "the bees will die!" because the protection is not there for winter, and that's how the old boys thought, but I was keen, young and wanted to experiment, so I began to make a bit of increase and the second hive was the one that I bought from Taylors in Welwyn Garden City. They are now no longer in business.

I had a nucleus in May. You had a choice: Taylors' Italians or Taylors' hybrids, so I thought "I'll have the hybrids." This nucleus came by British Rail. I put it into the second hive and it...struggled. I was very amateurish and didn't know why. It just managed to come up into the super and had about 6 combs with a bit of honey. I moved the bees off the garden to an allotment, and there was another beekeeper next door called Mr Roach who took me under his wing. I noticed these bees walking on the alighting board, falling on the ground, wings stuck up in funny directions. I thought, "there's something wrong here," but I didn't look in the colony. I gathered a matchbox of bees and sent them off to Taylors of Welwyn Garden City, and it came back via return post, "heavily infected with acarine." They gave me two bottles of Frow Mixture.

FROW a treatment for acarine

Steve: Of what, sorry?

David: Frow Mixture, F-R-O-W. A Lincolnshire beekeeper thought, "how can we cure this acarine?" The female got into the breathing spiracles of the bee and bred in there, and eventually suffocated the bee, that's why they couldn't fly. He thought the only thing was a fumigant that the bees would breathe in, that would kill the mites. He experimented, and he had this mixture called Frow mixture, and instead of supplying it commercially he gave the recipe to the beekeepers, and even Boots the Chemist – all of them. I've still got an original bottle somewhere of Boots' Frow mixture.

Steve: Wow. Did it work?

David: Oh yeah, it worked, if you followed the instructions. You had a pad on top of the feeder hole, and you put drops of this mixture on it and it fumigated the hive, but what I didn't realise by not removing the crown board, was that the colony was past it. It was only a small cluster of bees, and they died. All the frame tops were covered with dysentry and, you know, it was quite a blow to a young beekeeper.

But years later I told Adam about it and he said, "there you are – you have witnessed acarine disease that Dr Bailey at Rothamsted said never existed." It wiped thousands of colonies out in the British Isles. That's how he found the resistance to it by the Italian queens that were brought in; perfect observation – why have those colonies lived and the rest died? He kept records and saw it was Italians or Italian crosses that lived.

Breeding varroa resistant bees

Steve: Yeah

David: At the meeting here last week I spoke to Mike Brown from the bee unit. I said, "what about resistance to varroa?" He said nobody's doing work on it. I mean natural resistance.

Steve: Well, some people say they are. Some people say they don't treat their bees, so presumably any that live must have some resistance. Commercial beekeepers don't do that, they can't take the risk. Everyone I've spoken to treats their bees, because they see what happens if you don't, it's not good is it?

David: No. Well, commercially it's disastrous.

Steve: Some people think there's a type of bee that nibbles the legs off the varroa as part of grooming behaviour.

David: There's all sorts of theories.

Steve: In the end, if you want to keep bees, and you want them alive, you have to treat them really, don't you?

David: Of course you do.

I think it was Clive de Bruyn who said, tongue in cheek, when varroa was found in 1992 in England, that it had been here before along the East coast, and if we did nothing, what would happen? Well, of course it would have wiped out thousands of colonies. When acarine wiped out thousands the government imported skeps of bees from Holland. They came over on the sailing barges when the Channel was calm, and they came into the port of Hull. There is photographic evidence, I saw a photograph, it was quite a large one, of these barges stacked up with skeps of the Dutch heath bee, a noted swarming bee.

When I was gamekeeping I can remember those Dutch bees. I knew a beekeeper that had them. Swarms of bees used to hang from this tree like bunches of grapes, and when you looked at them they were those brown heath bees, and their crosses.

Steve: Carniolans are meant to be like that aren't they?

David: Well, I knew a beekeeper from Barnsley who would re-queen every year because the queens were so cheap, the carniolans, but they were very reluctant to swarm because they had been selectively bred.

Adam was working on this in his old age. He knew one of the African strains...you know he went to Africa?

Steve: The Monticola bee?

David: Yes, the Monticola.

Steve: They are up in the hills somewhere aren't they?

David: That's right.

Steve: Was it Ethiopia?

David: I'm not certain, but he went over there with a friend of mine, it cost him £2,000. They had to go right up to the edge of the tree line, where the locals kept them in log hives. They brought some queens back.

Pete Donavan sent me a couple of queens some years afterwards, that were Monticola crossed with Buckfast. Now, the African bee was supposed to be quite a vicious bee, but I put these two in hives to the side of the house, and they became immense colonies. When you worked on them they used to boil up, but they weren't vicious. You'd work on them in the daytime when the neighbours were out, and by the evening you could walk past and they wouldn't bother you. They got really tall, I had to prop them up. You couldn't keep them pure though, and I still treated.

Our bee hatches at 21 days but the Monticola hatches a little bit earlier, so the female mite couldn't come out with two or three daughters, maybe just one, something like that. That's the theory.

Steve: Did that go anywhere?

David: I don't think so.

Steve: Maybe over time they would naturally evolve that way. But they haven't enough time, have they?

David: No. The Asian bee that was the host of the varroa mite, nobody knows how many hundreds of years they have lived together compatibly. But it's quite rare for a parasite of any type to kill its host off.

Steve: Yeah, that doesn't make sense does it?

Revenue streams at Buckfast Abbey

With the hives at Buckfast was the income mostly from honey, and was making money even important? Was it honey plus queens or…

David: Mainly from honey.

Steve: Was it profitable?

David: I think so. In the early days it certainly was. Adam did rear a lot of queens. When acarine came along they went down to sixteen colonies I think, but with his expertise of beekeeping he was back to a hundred colonies within a season, by nuclei. He used to send nuclei and queens out, to help re-stock. He said the problem was that people would send their cash in for queens, but it all depends on the mating with weather and that sort of thing, and they'd be writing to him saying, "where's my queen?" and it was more trouble than a little.

Being a monk, and his way of life, you know, even when I was there he used to spend all of his Sundays answering, "how can I stop my bees swarming?" and all like that, and Adam said, right from the beginning, he said that beekeeping is simple. He said it's the beekeepers that make it complicated with all their different systems of management. He didn't have different systems, he had a colony of bees. The most important thing in there is the queen. He said about queens, all queens even good ones, "first year OK, second year OK, third year, if they start running out of queen substance they are liable to swarm," they're getting old you see, they've been laying 2,000 eggs a day.

There were different bees in the apiary, and some bees would want to swarm, and when you looked in the record book, and looked at the queen number, they were all sister queens. I'll tell you, all the time, it was pure observation you see.

Steve: Yes.

What about this; when you look on beekeeping forums nowadays, there's a lot of nonsense written but also a lot of sense too.

David: Yes.

Insulation and ventilation

Steve: One of the obsessions seems to be about insulation. What are your views on insulating hives, not just from Brother Adam's time but also from your years as a bee inspector? Some people think polystyrene hives are better and some put insulation under the roof…

David: It was like those old boys who said that bees will die in National hives, and they didn't. If the colony is healthy, and they've got plenty of food in there, and it's got a good roof, and they're off the ground (because cold comes up from the ground) – not too high off the ground, just so that when you work them you don't have to bend down, otherwise you get beekeepers back, they'll do well. In nature they are up high in a tree, sometimes they go low, but mainly they are up. And if the entrance is too big they'll close it with propolis and wax.

Going back to Buckfast, they had a crown board on, and a feed hole on top which had a wooden plug in it, and the board was varnished. The only trouble with it being varnished is condensation when the bees start breeding in spring, but it wasn't too detrimental.

Steve: So he didn't do anything about it? [the condensation]

David: No. In the States I don't know what their roofs are like, but Adams were very light because they were being opened up every week in the summer, so everything was done for lightness.

Steve: In parts of America they seem to have solid floor, not the mesh, but they also have an upper entrance. They seem to think that helps. I guess they must know what works for them.

David: Exactly. In Canada they used to winter them in cellars, or pack them in groups and stack them for the winter months. Adam did that with his half standard British National nuclei. He'd winter them at Ashburton apiary, in wintering boxes, with one queen and one entrance. There were four hives in one box, packed with shavings.

In the spring they used to find the queens in each one and then split them up into four nucs. Adam would equalise, a comb of honey, a comb of brood, comb of bees and so on, and the queens were put in cages to be used for re-queening or sent to his friends. Then you put them in the van, and they breathed through a hole at the bottom of each nuc covered with mesh. The top was sealed on and they put them on strips of wood on the van. We could get 25 in a load you see.

Steve: OK

David: We'd take them up to Dartmoor, about 12 miles away, put them on the stands, open the catches on the entrance, and the bees came out and recognised they had a new home, because they'd been moved. Probably a day later he'd go around and put a queen cell in each one.

Steve: He had it worked out, didn't he?

David: Down to a fine art.

After Buckfast Abbey

Steve: After you left there what did you do next?

David: I left there at almost Christmas time, and Buckfast kept paying my wages [laughs]. I think they had started a pension scheme and were paying me back what I'd put in. I'd heard about the bee inspector, Ivy Jakes I think her name was, she was formidable – worked for the ministry inspecting for European and American foul brood – she was retiring. Another man took the job on, a retired miner, and he thought it was just going around on nice sunny days looking at bees! But it wasn't so, and he used to get lost a lot, I think he wore several cars out and things like that.

Steve: Right

David: Then he was retiring, and the job came up. It was advertised in the "Notts Beekeepers" and I went for an interview and got the job. In those days it was only six months employment, April until the end of September, and in the Ministry they could put you on another job – I used to work on subsidies, hill cows and hill sheep, and then when 1st April came round I'd be back on bee duties.

I did work one year on the setting up of Rushcliffe Country Park. I could have changed jobs then. I could have worked for the County Council as a ranger on the country park, but I liked the Ministry work better. I kept on for quite a number of years, until 1992 when the budget was taken away – I went in one day and all my stuff was thrown in a pile on the floor [laughs]. It changed over to the Central Science Lab, and they had more equipment and it was a bit more organised.

They started having Regional bee inspectors, and I applied for a job and managed to get it.

Steve: I guess it was good job security and pension, being a civil servant?

David: Yes, that's right. It was all Official Secrets Act and things like that. It was good, I liked it. Of course, it gave me the opportunity to meet loads and loads of beekeepers of all different descriptions; commercial boys, hobbyists with one or two hives.

Steve: You were a bee inspector for a long time then?

David: A long time.

Steve: So you know everyone I imagine. You know everyone in the world of bees?

David: Well, quite a lot of people. At Buckfast we used to get people coming from abroad, especially from Germany. They used to come at queen rearing time. There was one American who came, he had a big cigar, and he slapped me on the back and said, "this guy's the best!" That was quite something, for an American to say that.

I think he [Adam] was a man well before his time; a very down to earth practical man. He was awarded a Doctorate of Science afterwards, because he was doing work that nobody had done before.

How did Brother Adam raise queens?

Steve: On the way he used to raise queens...did he used to have a queenless starter colony?

David: I think the Americans have starter colonies and finishers, and they transfer the cell. What he did, he found out from his records that the first ten days in June were the best for mating queens on Dartmoor, and we worked around that. He said that one year in every so many it will backfire on you, and it did one year whilst I was there, but usually it was the best time.

On the right hand side of the home apiary he used to devote 10 to 12 hives for raising cells. He would go out to the out apiaries with floorboards and brood boxes, and from the strongest colonies he would take combs of sealed brood. He would put 12 combs in a box from all different hives. People say, "don't they fight?!" but no, a puff of smoke and they would amalgamate fine.

We used to come back to the home apiary and make double brood colonies adding maybe six of the combs [from out apiaries] in the top box and six in the bottom.

Steve: Double Dadant brood

David: Double brood

Steve: That's a lot of brood

David: That is a lot of brood [laughs]. He used to select combs that would hatch in a certain number of days. On the morning of the graft Brother Pascal would lift the hive off the stand and put it to the side, put a new floorboard on the stand and put the top brood box onto the floor. It had no means of making queen cells because all the brood had hatched. He used to go through the colony, find the queen and put her on one side. Then he'd shake bees in front of the new hive, put the queen in the original hive, put a screen on and take it [the original hive] up to one of the out apiaries well away from Buckfast, usually Dean Prior. He'd leave a gap in the middle, and put a feeder on without any syrup.

He was brilliant at getting about 6 breeder queens to lay on the same day, so he knew, third day, he could select 12 hours old larvae. He had a special frame made. We used to get royal jelly – there was always a colony wanting to push cells up – so you got fresh royal jelly and diluted it with pure distilled water. The queen rearing house temperature was brought up to the temperature of the inside of a beehive, and a kettle was kept boiling to provide the humidity.

Steve: OK

David: There was good light at a window and an enamel table. Brother Pascal would put a drop of this royal jelly into the cells – wax artificial cells that were made over the winter or early spring – and they were put in the warming cabinet to bring the temperature of the cell up to what it would be in an actual hive.

Steve: Right

David: He'd take the comb out of the breeder colony, if the queen was on it pick her off and put her back in the colony, brush the bees off (not shake), hold it up to the light and look for a section where the larvae were all 12 hours old. Then he'd cut it out with a knife and put the comb back in the hive.

Steve: OK

David: Then he'd get a hot knife and cut the section of comb down to within 1/8th of an inch of the midrib. So he'd basically got a bit of foundation in his hand, and with all the light you could see the larvae.

Steve: That's good. I think even I might be able to manage that [laughing]

David: You can see them, you're not looking down on a piece of old comb. Then it was just pick them up with a grafting tool and refloat, and when he'd done a bar, cover it over with a damp cloth and put it on the table. He'd do three bars, so that's 60 cells, then it would be taken carefully in a cloth to Brother Adam who was waiting anxiously like a man waiting for a child. He put the frame in the gap in the cell builder colony with a finger space on each side so the bees had plenty of room to cluster around the cells.

He worked on a percentage. I think it was about 80%. Some years he'd get nearly 100% but other years less. And then the grafting would be over. We started at about ten o'clock in the morning and by lunchtime it was all done.

One day we did a graft at Ashburton apiary, with a board on the steering wheel, and Brother Pascal said, "here, you have ago," and it was accepted OK. Another day at the apiary I think Brother Pascal and Brother Adam had had a bit of an argument. Pascal came out wiping his hands on his apron. He just pointed to me and said, "you'll have to do it." That's how I got into grafting that morning, and I just did what I'd been taught and got fantastic acceptance

– Adam was smiling from ear to ear, "would you like some queens Mr Kemp?" I took the cells home wrapped in cotton wool in my shirt, cycled home, made my nucs up and popped the queen cells in. They were mature cells.

Steve: Great

David: Fantastic

Starving bees!

Steve: So you said there was a time when he got the timing wrong, or something?

David: Brother Adam was going to go abroad to Europe and he was in a rush to get the cells into the nucs, and I think he had been neglecting the production colonies a bit, and the weather was lousy. I think he left on the Friday and on Saturday morning I went down to the apiary, and there was a young chap called Michael, he was a chef in Mayfair and had come along to learn beekeeping, and as we were going down I looked across and could see drone larvae being thrown out of a hive. Then I saw another one, and another. I got the apiary book – sister queens.

I went over, picked the super up – nothing in it. There was not a stitch of food in the hive and the bees were chucking the drone brood out. I can remember Michael's look and he said "what's the matter Mr Kemp?" I said, 'go and have a look at them, where you can see drone brood being thrown out.' Even the best colonies hadn't got much.

I went straight down to the Abbey, got on the phone and rang Tate & Lyle in London. I said, "Buckfast calling, good morning London!" and he said, "Good morning Buckfast, what can we do?" I said that we needed four tons of sugar. "It'll be there tomorrow."

They'd got a depot on Dartmoor, and it turned up in a lorry, all in 100 weight bags. Father Richard came out and said, "What's happening, what's happening?" I said, "The bees are starving."

When you get that you must act immediately otherwise you're going to be in trouble. We went round and gave them, you know, a couple of litres of syrup, because you don't know when the weather's going to break again. But it was lousy for about a month, Dartmoor shrouded in mist...Adam came back, and the first thing he said was, "What's Sherberton like?" I said, "What's the weather been like in Europe?" and he said it had been just the same. I said, "we've got 150 mated out of 600."

If they're not mated within a certain time they'll start laying drone brood and that's the end of it, so I had to get some cell builders built up ready for another batch.

Steve: Do you remember what year that was?

David: Let me think. It was '70 something, I know it was a bad year.

[checking the Monthly Weather Report from the Met Office shows June 1972 was a bad year with 10 days of rain in the South West and mean temperature 3.1°C below the norm and sunshine 61% of the norm]

https://www.metoffice.gov.uk/binaries/content/assets/mohippo/pdf/q/7/jun1972.pdf

When I went there in '64 I know '63 had been a bad summer. One year we didn't take them to the moors because the heather wasn't looking right and the weather was bad. But you get a feeling. These old farmers are just the same. There's one I know, he said, "It's going to snow." I said, "do you think so, Bob?" and he said, "You know it is, the wind's in the Derby Hole." It was years of living in a place. They get to know the climate you see?

Steve: Yeah

David: But that was a disastrous year. We got some queens…but that's why you must rear so many queens every year, because one year will come along when you can't get them.

How to introduce queens

Steve: OK, so introducing queens, I think I read somewhere that he would just put new queens straight into the colony and they would be accepted because they were laying queens, is that right?

David: He'd go up to Dartmoor with cages. If it was spring there would be a picnic basket with a copper water bottle underneath and blankets [with warm water in the copper bottle]. He'd take the queen out of the hive, clip one third of one wing – they weren't marked in those days because we all had decent eyesight – and pop her in the cage. You put about four attendant workers with her, maybe half a dozen if it was very cold, plug the cage up (he had his own special cage), plug it up with candy, of his own special mixture [laughs] and put them into an envelope with a number on. They would then go into this warming basket.

When we got the required number we were off to the out apiary. He would go up to the hive, put the roof cater cornered [diagonally] on the ones that needed re-queening, leave the queen in her cage on the crown board, and our job was to find the queen, and if it was still good it would go to Dr Harding or Rothamsted, and if it was no good it was killed. But you see, we often took the old queen back to the mating station to go back into the nucs, to keep them ticking over.

Steve: So it was introduced in a cage, it wasn't just running the queen in?

David: Yeah. It was a laying queen, only in the cage for a couple of hours, put straight in so the bees could eat the candy and release her. Within a week or less we'd go round checking to see if the queen was on the comb then quickly close up. You got a few that were balled or went missing.

There was one interesting thing. We had a very good mating season; they got mated very quickly and the weather was ideal on the moor and the nucs were getting big. When you looked there were all these beards of bees hanging out the front. Adam said we had to take some bees away because otherwise they would swarm. Very early in the morning we took a floorboard with maybe four supers, with nothing but empty frames in, and he'd go around flipping these bees off the front of the hives into this box. The queens weren't in them, they were in the hives.

He'd build these boxes up and filled them with bees and put a screen on. There were six of them, absolutely teeming with bees from all over, all different crosses mixed up. Every so often he would stop the van and sweep water across the screens with a brush, because they'd overheat you see.

Steve: Right

David: We'd get to the apiary, I remember getting to one apiary and putting one or two of these hives on a stretcher and carrying them in. He'd look at his book, and if his notes showed any colonies needed help he'd go up to it, put a bit of smoke across the entrance, take the roof and crown board off, then take a frame or two of these bees we'd brought in and shake them into the super.

We were going along quite happily, then I looked at one hive that Adam had done, and the bees were fighting one another down the alighting board. I said to him, "Number 188 Brother, they are fighting." "Oh," he said, "the Greeks and the Turks never could agree!" [laughs]. He had a shrewd smile on his face. He shook one race into the other and they were scrapping right through the hive and coming out of the entrance. He could see the amusing side of things.

Selecting apiary sites

Steve: What about selection of apiary sites, presumably they were corners of a farmers field or something?

David: Mainly old cider orchards, that were not used or neglected. He paid his rent with a 28lbs tin of honey or something like that. I can remember all the apiary names even now. Staverton was a bad one because you had to go through a gate and across two fields to get to it, and there was Dartington Woods there on a slope, it used to drain. We had the van stuck many a time in that field.

But all in groups of four, and in rows, generally two rows. [the hives]

At the end of the season he was already preparing for next season. The queen excluders would be taken back to the Abbey and scraped, and vaseline put on them, then stacked with a bit of paper between them. He'd go out, he'd be out on his own somewhere, and he'd take 40 hives, 40 queen excluders, 40 supers from the abbey and put them in the huts in the out apiaries, ready for next season.

Braula mites

Steve: So what happened with the wet supers after extraction?

David: He'd put them on the hives nearest to the hut, sometimes four per hive, just to get them cleaned, then they'd go in the hut. He never stored them wet. I never saw wax moth down there. I saw it when I got back up into the Midlands but there didn't seem to be a heavy load of wax moth down there. What they did have was braula mites, I don't know if you've heard of them?

Steve: Oh, yes, they are a bit like varroa aren't they?

David: They are a bit like varroa but with less legs. I've seen them mating in the hive. They used to get on to the queen, I've seen the queen absolutely smothered in them. You could pop the queen in a matchbox and blow cigarette smoke in there and all the mites would fall off. Adam used to use Folbex strips, which was an acaricide from Switzerland. It was a fumigant. You'd light them in the hive and seal the entrance up. You could see smoke coming out of every crevice and the mites would drop off, he used to go around afterwards and stab them with his hive tool, he hated them.

Swarm prevention

Steve: What did you used to do about swarm prevention?

David: It was a weekly inspection. Some people say 10 days, but Adam said that if you leave it 10 days then it rains, and you have to leave it a couple more, you could be in trouble. So we would check them every week in the summer, two or three minutes per hive, and go through the apiary. A three-year-old queen would be more likely to start throwing cells, and some strains were more likely to do that. But then if a flow came along, a real good flow, many a time you could see a queen cell where the bees had taken the larva out and they were taking the royal jelly out, and you could guarantee that they'd packed up swarming.

Steve: For the whole season?

David: Yes, because time was going on, all the time.

Steve: It always seems that when they are stuck indoors in bad weather that's when they decide to make queen cells. I don't know if that's true or not but it seems that way to me.

David: Well it is, yes. They're all female and they get up to mischief when they're shut in the hive [laughs].

Steve: So if you come along and find some cells...

David: We used to pull 'em down, and sometimes they give up, sometimes they persist. He used to look in his records and if you've got a strain or so many hives in an apiary that were persisting, they're nearly always headed by sister queens. That would go down as a mark against the colony, you see.

Steve: Yeah

Breeding for variety not purity

David: Fantastic really. All those bees from all over the place – nobody else could see that – the information from all that variety of types of bees and crosses, and what they do...

Steve: And all that work...it's carried on now in Germany isn't it?

David: Denmark and Germany mainly.

Steve: So by now, I don't know if it works this way, but do they just continue to improve? So by now they must be even more productive, even more gentle.

David: Carniolan was in the Buckfast strain early because of its gentleness. A lot of the Italians, when they became very yellow, they used to turn everything, all the food, into brood. They'd starve quite quickly. But there was a strain that was not bright yellow, they were a browny colour. Buckfast drones; they used to call them leather coloured, they were dark but not black, they had these dark Italians in them.

The Buckfast strains were all as even as peas in a pod, there were no odd ones. In nature, remember she doesn't breed for purity, she breeds for variety. That's right away across. If you want a good dog you get a mongrel. You know, with breeding Alsatians they've had hip trouble and all sorts of things, because that's people who don't know what they're doing, and there's no fresh blood coming in. Nature breeds for variety in case something tragic comes along, and they don't all die then.

Steve: OK, so in your many years what's the worst you've seen in terms of pests or disease?

Worst times as a bee inspector

David: Well, I wasn't around for the acarine epidemic but I lost that hive to it when I was a boy. Winter of '63 was very bad. It was a prolonged winter and I had quite a few losses. Then after that I went to Buckfast.

Steve: They didn't have too much in the way of foul brood or anything did they?

David: Towards the end they did. It was a tragedy. When I was working there an odd colony would be disbanded, and Adam would sterilise the hive. I didn't know until I became a bee officer that Devon was the hottest spot for American foul brood in the country. It was tragic. As Adam got older he couldn't see, and they got a big outbreak of American, and the head of the bee unit said they wanted me to go down and help them. "If you don't want to go say so," he said, "it won't be held against you." I said I'd go.

We went down there and Peter [Donovan] was about but Brother Daniel was on the bees, and another chap called Roger, who had come from GCHQ and was a retired Major I think. We went into the office and there was a computer there which was new.

The first time we went around we took 96 samples...positive. The second inspection we took 19, the third one 9, and it was disastrous. Disastrous.

Steve: So what, you had to burn them?

David: Yes, we had to burn them all. They allowed them to sterilise the hives with caustic, boiling. It took a number of years to eradicate. The odd one kept turning up, but to my knowledge they never found any at the mating station, always the home yard.

Steve: So is that the worst you've seen?

David: Oh no. No, I've seen worse than that. Amongst commercial men. Too many colonies to look at, moving boxes of bees from one site to another pollinating. I've seen some disastrous situations, both American and European. People start off and they get far too many colonies, and if you get AFB it's straight destruction. EFB is shook swarms nowadays and treatment. The cost of it is immense.

Steve: Yes. I guess they're insured, but still.

David: One of them said he carried his own insurance. He was starting on 1st April when all his bees were gone from Derbyshire down to Kent orchards. They come back from the orchards, go up country pollinating soft fruits and finish up on the moors, then back home in October. We finished in September [bee inspecting] so we never saw his bees. He made a mistake once and left some in a Kent orchard. An inspector went and I'm not sure if they had both AFB and EFB or just one of them, but they had at least one.

In spring the Kent people asked if I was going to inspect his bees up in Derbyshire but I said that they'd gone already, down in Kent somewhere. He had a standstill notice on him but he broke it and got a fine.

But yes, outbreaks of foul brood...the thing is there's not the inspectors on the ground as there used to be. I used to have a team, when I first went into it there were about four part-timers, and they'd get a list of apiaries to visit and after they'd done them they were finished for that year. When I became Regional bee inspector I think I had about six or seven working for me.

We found foul brood in people who have got the National Diploma in Beekeeping (NDB), it doesn't mean a thing – you want a good pair of eyes and know what you're looking for. In Nottinghamshire they put so much on this education and passing these exams.

I remember when I was in Devon I went for an exam in beekeeping at the local college, and the county bee adviser ran the course. He looked at me and said, "What are you doing here?" because I was at Buckfast, so we sat there talking. There was one bloke from the Torbay area who showed me how to put a frame together. Well, I'd put thousands together at Buckfast but it wasn't quite the way I was supposed to do it, you see.

Steve: Oh

David: Too much education, you see. Too much education.[laughs]

Steve: Well, there's got to be a difference between what it looks like in a book and what it looks like in a hive.

David: Yes

What is the biggest threat to bees?

Steve: In terms of the single biggest threat...actually, all these people are going on about, "oh, we're doomed, the bees are dying," I know we can't be complacent but it seems to me that they are exaggerating. What are your views on the health of honey bees, because there are quite a few doomsayers out there?

David: There was a period in farming when spraying was really bad. It was very bad. There was a chemical, hostathion...it was when oilseed rape came in. There were thousands of acres and they were infested by pollen beetle, and the spray reps would go in and say to the farmer that they had to spray or lose the crop. One of them I think has a bit of nerve gas in, made to penetrate the chitinous covering of the beetle, and I had three colonies of bees almost wiped out with it. I made a big mistake. At the end of the day I called in to check on the hives and saw buckets of dead bees on the ground, and those flying in were being thrown out. I scooped up the dead bees with my hands and put them in a bucket. Within five minutes I got the worst headache I've had in my life.

Steve: Wow

David: It had penetrated my skin. I went down to the lake and washed my hands in the water. I learned a lesson there – never touch sprayed bees with bare hands.

When the ADAS gang took over we used to do a bit of recording for farming in the morning to do with bees. There was a meeting with all these young ADAS chaps there, and they were saying, "oh they are saying all these bees are dying – where's the evidence?" I put a paper on the table and threw this bucket of dead bees on. They all drew their chairs back. "There's the evidence."

David: That hostathion was produced in Germany. The German beekeepers had it bad, and they took it off the market, and sold their stocks to the English. There was some terrible stuff.

When you look around now...the Sandland area used to be all ragwort and rabbits, as we used to say, and after the war there were subsidies available to plough up and farm this land. That's when chemical farming came in, you see. A lot of that land now is like a medium just for growing stuff in.

With oilseed rape they used to burn all the straw and now they can't, so the rape is not so high. These plant breeders are brilliant, they can work wonders.

You get barley crops now, huge beautiful crops of barley with short straw. In the old paintings by Constable, years ago, they were cutting straw with scythes and it was up here [gesture to show height of straw].

Steve: Yeah

David: But it can be very dangerous. Remember those contaminated eggs that came over from the Netherlands? [Fipronil in eggs, 2017]

And cattle. I had a friend who was a farmer and also a beekeeper. He went to a friend of his, another farmer, who asked him to help carry this cattle feed in bags because it was going to rain. On the last bag this chap got hold of it, and his fingers went into it, and he said, "Good God, what are you feeding your cattle?!" Animal derivatives, you see? He took it up with the Ministry but didn't get anywhere with it. It was feeding herbivores with animal waste. You can't do that, it's totally unnatural.

Steve: What about those neonicotinoids?

David: Well that's still ongoing isn't it?

Steve: People I have spoken to about it, beekeepers, say it isn't a problem but those who study it say it is a problem.

David: Yes. I don't know the results of it, but I'm very sceptical about these things. The trouble is it's food, you see. I don't think England could produce enough to feed its population.

There was a vicar, and somebody said he'd got bees so I called in to see him one day. He said they weren't his bees, they belonged to his son who had left for university and lumbered him with them. He said that they had swarmed the other day, so I said, "what did you do with it?" He told me he put it in a cardboard box and dumped it up a lane. I said, "Oh dear, that's not right." I went to have a look, and there was this swarm of bees in a box under a sign which said "No Litter" – we had a laugh about it. He asked me in for a coffee, this was a long time ago, and he said "You know David, democracy hangs by a cotton thread. Close the supermarket doors and you'll see fighting in the streets."

He's right. Everything runs smooth while it runs smooth, doesn't it? And food waste nowadays, there's an immense amount of waste.

What would I put right if I could?

Steve: So if you could wave a magic wand and change anything what do you think it would be?

David: It's like the people who go on about the black bee. One man said to me that Adam was the worst thing that happened to this country. I said, "did you know him, did you see him?" "No" Well how can you say that? He was a man before his time.

If you could turn back to the 1700s when England was profuse with wild flowers…even in my lifetime – the herbs – the sides of the road were smothered in wild stuff. Then the fumes from traffic and sprays came and that sort of thing. The forage is just not there now. There's areas in Nottinghamshire, when the oilseed rape's finished there's nothing for bees, maybe a few limes around a church.

Those three years I spent on forestry…I wish now I'd stayed in forestry, because it's the most satisfying job you can have. I've got a landowner now, who said to a gunsmith friend of mine, "I wish that young man would come out who planted my field up with acorns." I must go and see him.

There was a wood I helped to plant. We planted oaks and beech, a few ash, a few Norway spruce. The land was heavy and suitable for hardwoods. That was Ladywood, and then there was a field next to it, which was a rough coarse field which the Forestry Commission had fenced off. We had to catch the rabbits, because they didn't have tree protectors in those days, and we planted it with acorns. We got the rows straight by lining up three hazel sticks with the tops peeled off, and moving them along as we placed the acorns, it was like sowing, going up and down. They were planted about a metre apart just to get the saplings to go up, you see, then you keep thinning out and thinning out until you are left with the best ones. They are going to be there for a hundred or two hundred years…

Steve: Yeah, lovely trees.

David: And Sherwood Forest used to be, what shall we say, loads of oaks, but then the coal mines came. My grandfather was a lead miner over in Derbyshire. Then the lead mines were worked out so they came over here to sink the coal mines. That took the water away. Rainworth Water was a wide shallow stream with trout in it and all sorts of things. It got shrunk right down. But the mines have closed now so the water levels will come back again over time.

Steve: So there's not so many trees in Sherwood Forest now then?

David: No. There was supposed to be a new one being planted. I asked some people from the university about it and they said the delay was because they can't decide what species to plant! I thought, we used to plant acres like lightning. You just have to get on with the job, you know? You find out about the soils, what's likely to grow. The pine forests around here were grown for pit crops, but the mines have gone so there won't be a market for it now.

Think of all those great houses and their estates, like Welbeck and Clumber, we wouldn't have the benefit of those trees if the previous owners hadn't planted them many years ago. The beautiful lime tree avenue at Clumber Park is magnificent.

Favourite books

Steve: I'm sure.

What about books? Are there any books that you have read over the years, old or new, that you have loved? Bee related, of course.

David: I'm still re-reading Adam's book, "In Search of the Best Strains of Honeybees". When I was there he gave me a copy and he wrote in it, "for services rendered." Also, Colin Weightman, who was a great friend from Northumberland who died a few years ago. He was a down to earth beekeeper, was Colin. I used to go up to Newcastle, perhaps to give a presentation, and I'd go to Colin's and turn into the farm where he lived, and he used to go, "spot on time, like Brother Adam." He was a big friend of Adam's and had a lot of correspondence with him, but he was a farmer. He knew his cattle. He went abroad doing photographs for his book, which is called "Border Bees," by Colin Weightman.

Have you read Adam's books?

Steve: I've read "Beekeeping at Buckfast Abbey." There was one which got very technical about genetics which got me a bit lost

David: "Breeding the Honeybee"

Steve: Probably. I probably should study it, really.

David: Yes.

Steve: It's something you can't just sit down and read, you have to come back to it.

David: You look at sheep and pigs and everything now, they are all hybrids or crosses. Racehorses are the Ferraris of horses. When you have such stock you have to manage them properly. You don't put a donkey in the Derby. Those stables at Newmarket; you could eat your dinner off the floor there. Those animals are looked after brilliantly, and then, when it's entered in a race you've got to put a good jockey on it's back, who knows his stuff. It follows all the way through. Selective breeding has been going on for years.

Black bees

One day one of the chaps who worked for me in Yorkshire put some bees into a greenhouse for pollination.

Steve: Yeah.

David: He said, "Come and have a look." It's the first time I'd seen hives in a greenhouse with supers of cabbage honey on them. It was shut in, and the plants had to be harvested by hand. This plant breeder there took us out for lunch and told us how they were working with the French and the Belgians to produce hybrid seed, for better crops and so forth.

He looked at me and said "Is there any beard and sandal brigade in beekeeping?!" I knew what he meant, and said "You're right there, it's all the 'black bee' people." OK, if you've got the British black bee and you can turn the clock back so we have fields of clover, then fine, but you can't. You've got to keep moving forwards.

Steve: It's strange isn't it? Beekeepers should all get on shouldn't they? But it seems to be human nature to form tribes and fight each other. People with black bees arguing with people with Buckfast or Italian bees, I don't understand it really.

David: Yes. I don't know if it's the same in America but the British are…I mean they're still designing beehives for goodness sake, you know? Adam said that progress in future would come from better strains of bee, not hive designs. At one time we were attached to Europe until the Channel broke through. Probably what they call the "native bee" was isolated when the Channel broke through and adapted to its environment at the time, and it was quite good. But not in modern production of honey.

Different sorts of hives

Steve: I noticed when I was at Mike Palmers place that he's big on Brother Adam. He's big on queen rearing and has lots of nucs that he over winters; four frame langstroth nucs which are placed side by side so the cluster forms at the joining walls. His nucs don't have the dovetail joints because he says where he is they rot too quickly, so his joint is the one down the whole length of the side.

David: A rebate joint [rabbet joint in the USA]

Steve: So little things like that do make a difference over time.

David: Of course.

R.O.B Manley's are interesting books to read. "Beekeeping in Britain" and "Honey Production in the British Isles." He had 2,000 of his own hives. Another one was Alec Gale, who was a great friend of Adam's. He never wrote a book but one of his chaps wrote one about him, "A Man and His Bees" by D A Clements. He also kept 2,000 hives. Once I was in the honey show in London I met a chap who I knew had worked for Manley. I said "Didn't Manley used to use some Buckfast queens?" He said yes.

Also Adam would send Alec Gale queens, because if you can send queens to a man with 2,000 hives (Adam only had 320) you'd be getting a better feedback. Gale would not tell him any lies, so if they were no good he'd say "They're no good, Brother." I think Alec was suppose to go into the printing industry, but he didn't. Adam said he only slept about 4 hours a night but he always put a lot of time in to seeing his children. Then his eyesight failed him. He used to go in the farm in the morning and set the workers off, then down to the bee department and he'd drive this truck with his fading eyesight terrifying everybody. Adam said he use to live on lettuce sandwiches.

Athole Kirkwood was another with a couple of thousand hives. He was in Scotland and used Smith hives, the ones with the lugs cut down. He pondered over hive types for a while and realised that it was there all along, the Langstroth hive, and apparently he ordered a thousand from Thornes and knocked them all together one winter, and moved over to Langstroth hives. In fact, I've got two of his brood boxes which say "Heather Hills Honey Farm" on them.

Steve: I like Langstroth hives

David: Yes, why not? But a hive is just a box.

Plans for the future

Steve: What about plans for the future?

David: Well I'll be eighty. I'm just freewheeling now, and enjoying it, you know? Listening to other beekeepers. I'm very much into the countryside. I still do a bit of shooting, not much.

Steve: What do you shoot, birds?

David: Er, yeah, I've done a bit of ground game shooting but I don't...it's like Sir Peter Scott, he was a great wildfowler, with long barrelled guns, I think he drove motor torpedo boats during the last war, but the older he got, the more he became interested in painting. He set up the Wildfowl & Wetlands Trust. You don't want to take life anymore.

I shoot rats, because rats invade my bird table. I don't like poisoning rats because it's a slow death. I've got Brother Adam's gun! Can you imagine that, a German monk with a gun? How he got it back from Germany I don't know, probably in his habit. It's a German 9mm smooth bore, not rifled. It's on my shotgun certificate and it's a great gun for ratting. Very small bore and not a big range. In one year I shot over a hundred rats off a bird table, out of my bedroom window.

Steve: Wow. Is that early in the morning or something?

David: They feed at 4 o'clock in the afternoon. If you put some food out then they'll soon turn up. The rats could crawl up a nearby bush and get onto the bird table. I used to get the bedroom window open and just wait. Sometimes you could get two with one shot. The trouble is the bird table began to look like it had been blitzed!

Rats are very astute creatures. There's loads of them in Nottingham. They say you are never far from a rat.

Steve: Do you do much speaking nowadays?

David: I used to do a lot, but with traffic and distances now I don't get out very much. I used to give loads of presentations all over my region, which was Notts, Derby, Leicester, the whole of Lincolnshire and Yorkshire. It was a massive area. I also used to go up to Newcastle, out to the Wirral in Cheshire, Gloucestershire over the Cotswolds. The beekeeping meeting would open up about 7pm and afterwards I'd drive home through the night, because it was quiet on the roads. If you stayed overnight you were caught in the rush the next morning.

Steve: So you hang out with your RAF mates don't you?

David: Yes, they'll be here tonight.

Steve: Were you a pilot?

David: Oh no, I was a cook, for two years. I wanted to be an armourer, you know, re-arming planes when they came in after their flights, but they wouldn't train you unless you signed on for the Air Force. I thought two years [National Service] was enough. If you signed on it was for nine years.

Steve: That would have been a whole different story, wouldn't it?

David: Yes, entirely, because I wouldn't have met the people I met. I could have gone back to forestry. But I was glad to go to gamekeeping to see the end of an era. The old Duke, the one I worked for used to be the Marquess of

Titchfield, but his father, realised after the 1st World War that they wouldn't be able to continue to live in the abbey. The place was coal fired, you see, with servants running up and down with buckets of coal...look what the price of coal is now. So he had a house built in Welbeck Park that was more manageable. They could see into the future and make plans. I like the House of Lords because they have got wisdom.

So, what's it all about?

Steve: What, after all these years of working with bees, with Brother Adam and all those other beekeepers, do you think beekeeping is all about in a nutshell?

David: I think it's no use thinking, "Oh, the bees are under stress, let's keep bees," which is happening a lot now. You've really got to have your heart and soul in it. I think a lot of people in the towns have a yearning for the countryside but unfortunately they live in the big cities. Suburbia is now very good for beekeeping with all these gardens and things like that.

When you get into hilly ground, where it's not suitable so much for modern farming with combines and great big fields, that's better for honey production.

Steve: Yeah, OK, so you would say that you have to love it.

David: Yes. I mean, there was no-one in our family who was a beekeeper. You've got to have an affinity with nature. I always get on well with dogs, and cats. I used to have two gun dogs, two labradors, and I've got a cat now who's on his last legs, he's 17 years old. People blame cats for killing birds but grey squirrels and magpies will kill far more birds. Magpies will take eggs out of nests and young birds, as will grey squirrels. And what they call the raptors, sparrow hawks, there's hardly any sparrows left now.

On the grouse moors – everybody says we should close the grouse moors and stop shooting but if we stop shooting there will be no grouse in a few years, because the raptors would take the grouse away and the heather would grow rank (because burning keeps it down) and, you know, it would change things. Everything in moderation.

I was a partridge keeper and all the ground vermin had to be trapped and killed. I've seen some of the lists from the big estates in Norfolk. They kill thousands of stoats and weasels, every conceivable thing. When I was with the forestry commission the land had been neglected because of two world wars, and these woods went wild. If you've ever read Wind in the Willows, there's a section on the wild wood, well that's what those woods were. They were teaming with stoats, weasels, rabbits...We used to sit around the fire and then we'd hear a rabbit screaming, because a stoat was near, and the lads used to get up from the fire and make their way towards the sound. I've seen rabbits trembling, with a stoat dancing around, mesmerising it. Working in woods like that is fantastic.

Steve: What's your favourite type of honey? Do you like heather honey by the way?

David: Yes. The heather plant Calluna vulgaris when grown on Dartmore, which is granite is different to other areas. When you come up to Derbyshire it's gritstone undersoil, then over to Yorkshire and they are peat moors. The Dartmoor one was a very rich dark port wine colour and very gelatinous, and yep the Derbyshire one is OK. When you get up to Scotland on the Highlands, I've seen their honey in Waitrose, and very often it's contaminated with clover or something else, but if you can get it pure then it's quite nice.

But clover honey is...

Steve: Clover honey is the one for you

David: Yes. Clover first, then heather and then lime honey [linden].

Good years and bad years

Steve: OK, I don't know if there is an answer to this but...you know you can have good years and bad years – is it all about the weather? You get years where the bees seem to be more swarmy, good years for honey and bad, is everyone just floundering around hoping, or is there some pattern?

David: As a beekeeper you look after your colonies and get them up to strength for the time when you expect there to be a honey flow, but the rest is out of your hands. Certain trees will have a good crop one year and nothing the next. Nothing is plain sailing.

Steve: So we can't predict these things?

David: No, you can't predict. You can take your bees up to the moors and it will either be a good year or a bad one, you just don't know. Sometimes they'll fill the brood box and nothing else, sometimes there are stacks of full supers and they're falling over. You don't need bright sun early, for good honey flows you need mist in the morning, and the mist gets driven off by the sun, then the plants produce nectar and it pours in. Brother Adam said to me, "Even the gate posts will give you honey in a good year!"

You can feel it when you are a good beekeeper. They used to say that some of the big beekeepers in the States would take to their beds when the weather was bad, with bad backs or whatever, and suddenly the weather breaks and they'd be there, straight away. You can feel it. You can feel it in your bones.

Steve: So about your photographs, are you going to let me have some on a memory stick?

David: Yes, my friend Andy Wattam is going to help sort all that out. What kind of thing are you looking for?

Steve: Maybe some shots of you and Brother Adam at Buckfast and at the mating station? Maybe something that hasn't been seen in print before.

David: I don't think I have any of me with Brother Adam in the same shot. I have some of me at the home apiary, which was a beautiful place. There were even begonias along the pathways – Adam and I used to dig them before Christmas, and when we were taking a break we'd sit there with a monistic bowl and a bottle of 17 year old mede. After you've had a pint of that you can shift a ton of soil without feeling it, until you get home and have a bath.

He kept a wonderful herbaceous border with delphiniums and lilies, and all along the wall he had specimen daffodils. Then one day we took down the hedge and replaced it with a wooden fence, and we planted acacia trees on the right hand side. All the leaves from the lime trees around the apiary were raked into a heap. We would dig up the topsoil from a flower bed, place it on a plastic sheet, then shovel leaves into the hole before replacing the soil on top. It raised the beds and improved the soil.

In the summer going up onto Dartmoor, working on Saturday morning, you could feel Adam take his foot off the accelerator as he was driving, and I knew what was coming: "What are you doing this afternoon?" meaning are you able to work this afternoon.

He'd go to the Abbey kitchens and collect some scones made by Brother Andrew, and going up to Sherberton he'd stop and go across to this farm which had a notice up selling Devon clotted cream. He knew that the feeders on the nucs were full of heather honey. We used to sit down eating warm heather honey straight out of the beehive with these cream scones, and he'd say, "That's the way to eat heather honey."

Steve: I bet it is. Was that with the comb as well?

David: Yes. You'd crush the comb up.

Steve: He had a little mischievous side to him didn't he?

David: Yes. He was a very down to earth and sincere bloke.

Steve: What about those photos of yours?

David: What I'll do is call Andy and I'll get him to put a good selection on the memory stick. They'll all be Buckfast. They'll be apiaries...do you want the ones with wax melting?

Steve: Yeah, the more I've got the better. I probably won't be able to use all of them but it's interesting and it's quite important I think.

David: Oh yes, it is. I don't know where that lot I've got should go eventually. Maybe they should go back to Buckfast Abbey, I don't know. They'd put them in archives I suppose. I don't even know where the Ministry are now though, because Animal Health has taken over. Andrew resigned straight off, like that, because there's nobody in Animal Health who knew anything about bees. They call bees "food producing animals" but I've never seen an animal with six legs and four wings, have you? I don't know what they do at Rothamsted. For years that experimental station had bees, and the bee department up at Central Science, they've got quite a few hives up there. There's work being done but things change and get moved about. I think in America their establishments are more permanent, although Adam told me that it wasn't the same as what it used to be. Baton Rouge and these different places where they had bee departments, you know.

Moving bees

Steve: I know there are problems moving bees between America and Canada now. You have to jump through a load of hoops.

David: Yes

Steve: They are at least keeping bees in Canada over winter whereas they used to just...I think they used to kill them off each year

David: They used to kill them each year and then import the packages.

Steve: Now you couldn't do that because packages, apart from the fact that they're rubbish, you've got all that border control customs stuff. I don't suppose a bee from Texas or somewhere is going to do well in Canada.

David: No, that's right. The Americans – I was reading about it in Geographic, an old copy – I mean they just load hundreds of hives up don't they, and shift them about. It's like a big dose of foul brood we had once; a chap was loading them on trailers, and it puts bees under stress. When Adam moved his bees you went out in the afternoon and we could close forty hives down in half an hour. He took them to the moors with one super on, and when they were taking supers off if there were any combs that were unsealed he'd put them in that one super. He said that in moving they get excited and they consume food.

Steve: Was the moving done at night?

David: No. In the afternoon we put the screen on and he had two quarter inch rods, and there was a wing nut on the top. Short rods if it was a brood chamber and longer rods if there was a super on. They went down between the frame and the hive wall, down through a hole in the queen excluder, there was a little tin plate with a hole in it, and it went through the brood chamber into a counter sunk hole in the floor – one at the front and one at the back. On the floor was a brass plate, tapped with quarter inch Whitworth thread, and it went into there and you could tighten it up. The hives would be loaded, twenty on the bottom of this Cooperative lorry that we hired, then there would be a platform, with an air space, on top of the twenty. The next twenty went on top. It took three men; two to lift it up and the third to push it along into place on the lorry. I was generally on top with Mervyn, the driver, stacking hives up. You could bet your life that if some bees got loose Mervyn would be over the side [laughs].

On the centre of the load, on top, would be all the roofs, upside down, with the crown board and alighting board in. On the outside of those would be the stands, then it was roped down. If you've got hives loaded like that with screens on, not dumped on the lorry but put on gently, whether your journey was ten miles or a hundred it made no difference, because they were secure. When you arrived at your destination on the moor – it was only maybe 15 to 20 miles away – the first thing was ropes off, then hive stands. They'd be put down in groups. Next we used to take the roofs off, then the hives.

When they came back from the moor, whatever crop was on them was taken off earlier, so only the brood chambers came back. The hive numbers had to go in the apiary on exactly the same stand facing the exact same way as when they left, and Adam was a stickler for that. The way I got around it was to have a clip board with a plan of the apiary on it and the numbers of the hives. I made sure they were returned to the exact same place.

Steve: It wouldn't really have made much difference would it?

David: Well, it would because his hives were all numbered and it was all part of a whole system with records and such like. You couldn't have them mixed up because of the records.

Adam used to get me to come when these beekeeping associations would come around. I remember two chaps from the BBKA came down and one of them said "Oh, this is alright for Buckfast but we can't possibly do something like that." I thought "you fool!" You can run two hives on the Buckfast system as well as a hundred and two. It was a standard that they should have adopted, you see.

We had some Belgian beekeepers come down once. It took two of these big coaches to get them over Dartmoor to the mating station. That was a farce, because there were people on holiday there. One time we had to move this guys car off the road to let the coach go past. They were different; the foreign beekeepers were much more attentive than the English ones, they could see that it was good.

Best advice for finding queens

Steve: In general would you say it's better to breed your own queens or get them from your local queen breeder?

David: Yes, of course it is. This lass I know was selling Ged Marshall queens at £40 a time whereas the Danish ones were £65, and they're worth it, but there's this gap between them coming out of Denmark and arriving here. It's not like at Buckfast; straight out of the nucs and into the hives, and the time gap does affect them. The ideal would be to get the whole apiary with queens from a good source, then start breeding from them over here. Then they don't come at £40 each but a few pounds apiece. It's good to bring in new lines every so often from wherever.

I remember once getting hold of a cracking strain, they were Buckfast 339s, and I bred from them and gave some to a friend, Derek, who keeps 150 colonies of bees. These bees were so quiet. You could go there on a damp evening, no problem, and they produced as well. I said to him, "look Derek, this whole home apiary surrounded by this hedge have nothing but 339s in them."

His system of management was a brick. If the brick stood that way it meant one thing and if it laid flat another. I said "get some numbers on your hives and a book!" But it was drawing pins and bricks. I could go and bugger up his records just by shifting the bricks around! He just didn't grasp it. His son brought in some nucleus hives with all sorts of mongrel bees in.

Before Adam took those colonies up to the moor, if he'd got 12 colonies with all these cells in, he'd put all the cells in 2 colonies. You'd have a comb, then a frame of cells, then a comb, and so on. He was dead accurate on his timing so they didn't start hatching out on the way up there. But those 2 colonies had to be passed through a queen excluder because he didn't want any foreign drones in there. It's simple.

What happens with British beekeepers – they should buy all their equipment in the winter and put it together then – but they don't, they wait until spring and then the shelves are empty because everybody is going at them.

Respect for bees

Steve: One of the things I've noticed, and I've only been beekeeping for a few years and have about 10 colonies, but when I went to see this guy that did it every day for a living, I thought he would be quite rough with the bees, but he was really gentle. He would brush bees away from the edge of the box before putting another box down. I think actually, that somebody who works with bees every day – the last thing they want is to squash bees because they'll get stung more – so I learnt a lot just watching how gentle he was when working the bees.

David: Well Adam had a great respect for his bees. I know someone who used to suck swarms up through a vacuum cleaner. He thought it was the greatest thing since sliced bread. I said, "look, all those bees have got stomachs full of honey, because they are a swarm" and I said, "would you like somebody to put you in a centrifuge after you'd just eaten a big dinner?" You've got to have a feeling for these insects.

We used to occasionally get stung on the ends of our fingers but you could work for two days and not get a sting at all when the weather was right, it's just having a feel for them. Adam used to open them up and just put a little bit of smoke in front, just to let the guard bees know you're coming. The only fuel we burnt was wood shavings. When the apiary hut was closed for the winter there was a bin next to the door full of sacks of shavings. We used to tip them in there, all ready for the next spring.

Steve: Well, thanks very much for your time, David.

Acknowledgements

This all started with me becoming a beekeeper. For that, I have to thank Fay Hine for inviting me to a dinner party at which I met Richard and Pamela Winward. They let me have a place on their farm to keep bees. They have always been very supportive of my beekeeping efforts, and I am incredibly grateful. I must also thank Graham Royle for his inspiration at the start of my beekeeping journey. He taught my "introduction to beekeeping" class on cold winter evenings in Wilmslow, and it was his bees that I first handled; after that, I was hooked.

Huge thanks must also go to the beekeepers that I interviewed for this book. They were generous with their time and patient with my occasionally incoherent lines of questioning. They are busy people and could easily have brushed me aside. In particular, I'm enormously thankful to Michael Palmer, who was my first interviewee. Once I had his interview in the bag, I had some credibility with the next people on my list. Michael and his wife Lesley looked after my daughter and I on our five days stay with them in Vermont.

Sadly, two impressive and interesting beekeepers that I met in New Zealand did not make it to the book. They are Lorraine Muldoon and Rae Butler; I'm sorry that I could not tell your stories this time. I thank you for your warm welcome and for letting me peek into your lives.

I am blessed with a loving wife, Elaine, and three fantastic children; Clíona, Alexander and Isla, who all make me proud. Thanks so much for being there and for believing in me. Thanks to my mum for her quiet, somewhat bemused interest, and my dad for his effusive and unstinting praise. Dad passed away before I could finish the project, but I do not doubt that he'd pat me on the back if he could.

I am delighted to have obtained some rare photographs for the book, by David Kemp, from his time at Buckfast Abbey in the 1960s. Thanks to Andy Wattam for sending the images to me in digital form.

Finally, thanks to anybody who reads this book for your interest. I wrote it for me really, so if anyone else enjoys it, I consider that a bonus. It was a beautiful life-changing experience; I travelled, met some very special people, and learned a great deal.

* 9 7 8 1 9 1 6 2 0 4 2 5 6 *
BV - #0022 - 240320 - C51 - 234/156/21 - PB - 9781916204256